Design Center
Stuttgart

Focus Energy 2006

Internationaler Designpreis
Baden-Württemberg 2006
und Mia Seeger Preis 2006
—
Baden-Württemberg
International Design Award 2006
and Mia Seeger Prize 2006
—

FOCUS

avedition

Inhalt

Kategorien des Wettbewerbs

Contents

Categories of the competition

Energy ... Wirtschaft

Der Internationale Designpreis Baden-Württemberg steht dieses Jahr unter dem Thema »Focus Energy«. Welche Rolle spielt Ihrer Meinung nach das Design beim Thema Energie?

Die industriellen Hochtechnologiebranchen Baden-Württembergs, zu denen der Maschinenbau, die Elektro- und Nachrichtentechnik sowie der Fahrzeugbau gehören, haben eine unmittelbare Beziehung zur Energie und damit zum Thema des diesjährigen Internationalen Designpreises.

Die Palette der diesjährigen Einreichungen reichte von einem Lernspiel zum Umgang mit der Energie, kleinen Windkraftanlagen, mobilen solarbetriebenen Stromversorgern, Solaranlagen über Pelletöfen und Wärmepumpen, die Erdwärme beziehungsweise Erdwärme aus dem Grundwasser nutzen, bis hin zu einem Konzeptfahrzeug, dem Mercedes-Benz Bionic Car.

Der Schlüssel zur Zukunftssicherung heißt Innovation. Design ist ein wichtiger Teil des Innovationsprozesses. Ob Informations-, Service-, Produkt-, Investitions- oder Konsumgüterdesign – im nationalen und internationalen Wettbewerb entscheidet das Design bei gleicher Ausgangslage über die Wettbewerbsfähigkeit.

Hochwertiges Design ist in Baden-Württemberg ein wichtiges Element der Standortqualität und ein bedeutender Bestandteil der Wirtschaftskultur. Daher überrascht nicht, dass Baden-Württemberg bei sehr vielen Wirtschaftsindikatoren bundesweit überdurchschnittlich abschneidet.

Wo liegen die Herausforderungen der künftigen Energiepolitik?

Erklärtes Ziel der Landesregierung ist die Verdoppelung des Anteils der erneuerbaren Energien bis zum Jahr 2010. Ausgehend von einem Anteil von 6 Prozent im Jahr 1997 soll der Wert bei der Stromerzeugung auf 12 Prozent gesteigert werden. Damit nicht genug: bis 2020 soll der Anteil bei 20 Prozent liegen.

Dies sind zugegebenermaßen sehr ehrgeizige aber auch wichtige Ziele hin zu einer nachhaltigen Energiewirtschaft. Erreicht werden diese Ziele durch überdurchschnittliche Wachstumsraten der Energieträger Biomasse, Geothermie, Photovoltaik, Wasserkraft und Windenergie.

Neben der Nutzung erneuerbarer Energien ist die rationelle Anwendung von Energie ein Kernanliegen der Landesregierung. Rund 20 Prozent unseres Energiebedarfs können mit vergleichsweise »einfacher« Vorgehensweise – Steigerung der Energieeffizienz in den privaten Haushalten, öffentlichen Einrichtungen, im Verkehr sowie bei Gewerbe und Industrie – eingespart werden. Bei einem weiteren Fünftel erfolgt die Einsparung durch intelligente technische Lösungen.

Design, und hier schließt sich der Kreis, kann einen wesentlichen Beitrag zur Akzeptanz erneuerbarer Energien und der Durchdringung der rationellen Energieverwendung leisten.

Energy … economy

Ernst Pfister, MdL
Wirtschaftsminister
des Landes Baden-Württemberg
–
Ernst Pfister, MP
Baden-Württemberg
Minister of Economic Affairs
–

The theme of the Baden-Württemberg International Design Award this year is »Focus Energy«. How relevant do you think design is to the subject of energy?

Baden-Württemberg's high-technology industries, which include machine construction, electrical engineering, telecommunications and vehicle construction, are all directly linked to energy and thus also to the theme of this year's International Design Award.

The products submitted this year ranged from educational toys for learning how to use energy, small wind turbines, mobile and fixed solar power generators, pellet-fired furnaces, heat pumps that use geothermal or groundwater heat, to a vehicle study, the Mercedes-Benz Bionic Car.

The key to securing the future is innovation. Design is an important part of the innovation process. Be it information, service, product, investment or consumer goods design – in national and international competition, given the same starting position, competitiveness is decided by design.

In Baden-Württemberg, high-quality design is an important factor in the quality of the location and a significant part of the state's economic culture. It is hence not surprising that very many economic indicators put Baden-Württemberg above the average for the Federal Republic of Germany as a whole.

Where are the challenges of future energy policy?

It is the declared aim of the State Government to double the proportion of renewable forms of energy by the year 2010. In power generation the proportion is to be increased to 12 per cent, from 6 per cent in 1997. But that is not all: by 2020 the proportion is to be 20 per cent.

These are admittedly very ambitious but important objectives on the way to a sustainable energy economy. They will be achieved by above-average growth rates for biomass, geothermal energy, photovoltaics, water power and wind energy.

Besides exploiting renewable forms of energy, the rational use of energy is one of the State Government's prime concerns. We can reduce our energy needs by about 20 per cent by comparatively »simple« methods – increasing energy efficiency in private households, public institutions, transport, and trade and industry. A further fifth can be economized with intelligent technological solutions.

Design – and here the wheel turns full circle – can make a major contribution to the acceptance of renewable forms of energy and to rational use of energy.

Energy … Design

Energie. Design. Sind das nicht zwei Paar Stiefel?

Das mag auf den allerersten Blick so aussehen. Doch ist Design ohne den Einsatz von Energie überhaupt vorstellbar? Ich meine: nein. Ein Produkt industriell herzustellen, ohne dafür Energie aufzuwenden, erscheint mir heute noch unmöglich. Bei stetig schwindenden Ressourcen und steigender Nachfrage sollten wir uns vielmehr fragen, welche Energieformen in Zukunft sicher sind – umweltverträglich, kostengünstig und allgemein zugänglich. Gleichzeitig gilt es, mit der vorhandenen Energie verantwortungsvoll und wirtschaftlich umzugehen. Darauf haben wir uns einzustellen – als Entwickler und Designer, als Hersteller und als Konsumenten.

Focus Energy. Diesmal also Ökodesign im Fokus?

Weit gefehlt. Ich halte diese geistige Schublade weder für hilfreich, noch für zeitgemäß. Unternehmer und Gestalter, die dauerhaft Erfolg haben wollen, sehen ökologische Aspekte im Designprozess, in Produktion, Vertrieb und Service längst als feste Größe. Der Umgang mit Energie gehört dazu. Energie und Design, das sind nicht zwei Paar Stiefel, sondern zwei Seiten derselben Medaille – untrennbar miteinander verbunden.

Mir fällt eine Reihe von Stichworten ein, die ohne weitere Erklärung auskommt: Welche Materialien werden verwendet, ihre spezifischen Eigenschaften, wie werden sie gewonnen oder hergestellt, transportiert, montiert. Das Gleiche gilt für das fertige Produkt. Außerdem auf unnötige Bauteile verzichten und kreative Energie sinnvoll nutzen.

Was meinen Sie damit?

Energie bestimmt auch die Beziehungsebene – und damit den Erfolg. Hersteller und Designer, beide sind gefordert. Wo fantasielose Zauderer und Bremser das Sagen haben, denen es schwer fällt, sich das Machbare vorzustellen, wird die meiste Energie verschwendet. Industrie und Handel, Marketing und Vertrieb dürfen sich nicht aus der Verantwortung stehlen. Wer von guter Gestaltung wirtschaftlich profitieren will, soll sich dazu bekennen. Ohne Umschweife.

Energy … design

Henning Horn
Leiter des Design Center Stuttgart
—
Henning Horn
Head of Design Center Stuttgart
—

Energy. Design. Is there any connection?

At first sight you might not think so. Yet… is design without consumption of energy even conceivable? I do not think it is. It seems to me that even today it would be impossible to manufacture a product without consuming any energy at all. In the light of dwindling resources and increasing demand, we would be better advised to ask ourselves which forms of energy we can rely on in the future, and which will be environmentally compatible, affordable and generally available. At the same time it is important to use the energy we have responsibly and economically. This is the situation we must adapt to—as developers, designers, manufacturers and consumers.

Focus Energy. So the focus this year is on eco-design?

Far from it. In my view, this intellectual compartmentalization is neither helpful nor in keeping with our times. Entrepreneurs and designers aiming for continuous success have long since regarded ecological factors as a constant in the design process, in manufacturing, sales and service. The use of energy is one of those factors. Energy and design are not unconnected, but rather two sides of the same coin—inseparably linked, in fact.

Several keywords occur to me that need no further explanation: What materials are used, what are their specific properties, how are they obtained or produced, transported, assembled? The same applies to the finished product. And: dispensing with superfluous components, and using creative energy wisely.

What do you mean by that?

Energy also dictates the level of the relationship—and thus the degree of success. Both manufacturers and designers face a challenge. Most energy is wasted where unimaginative ditherers and dampers are in charge who have difficulty envisaging what is feasible. Industry and commerce, marketing and sales must not shirk their responsibility. Those who want to reap the economic benefits of good design should declare their support for it without beating about the bush.

Drei Szenen am Stück
Three scenes in a row

Wolfgang Berger
Design Center Stuttgart
–

Personen: Dimitrij und Goldtblom (betagte Aufsichtskräfte); Raumpflegerin (stumm); Aufwärterin (Chefin eines Partyservice, mit mehreren Bedienerinnen); Hausmeister; Fahrer.

Bühnenbild: überdimensional vergrößerter Kaffeeautomat ›WMF Presto‹ (s. Abb.), in dessen Front eingefügt der Eingang mit zweiflügeliger Glastür, so daß man die Auftretenden herannahen sieht; auf der Bühne die vollständige Ausstellung »Focus Energy 2006«.

Obacht

–

Dimitrij und Goldtblom auf dem Sofa (s. vorliegender Katalog, S. 134) liegend.

DIMITRIJ Paß doch auf, wenn ich dir was sage.
GOLDTBLOM Entschuldigung.
DIMITRIJ Schon gut. Ich sagte: Keiner da.
GOLDTBLOM Wo?
DIMITRIJ *aufwallend* Im Zuschauerraum, du *–piep–*. Wo denn sonst? Und in der Ausstellung hier ist auch kein Mensch.
GOLDTBLOM *unbeirrt* Gesetzt, es wäre Einer da, was sollten wir dann tun.
DIMITRIJ Aufsicht führen.
GOLDTBLOM Gegen wen?
DIMITRIJ Paß mal Obacht. Wenn du hier scharfsinnig werden willst, dann häng ich den Vorgesetzten raus, und du drehst deine Runden, ohne dass da Einer da ist.
Peinliche Pause
GOLDTBLOM Weißt du noch?
DIMITRIJ Was? Wie wir gewartet haben?
GOLDTBLOM Nein, wie wir gefroren haben.
DIMITRIJ Gefroren und gewartet.
GOLDTBLOM Wir frieren nicht mehr. Es ist warm hier. Es wartet sich schöner, wenn geheizt ist.
DIMITRIJ Aber wir warten nicht mehr. Wir müssen aufpassen.
GOLDTBLOM Hier ist das ein und dasselbe. Wir sind Wärter.
DIMITRIJ Wir müssen sehen, daß den Exponaten keine Unbill widerfährt.
GOLDTBLOM *im Sprechgesang* »… der Herr hat sie gezähählet, daß ihm auch nicht eines fehehlet.«
DIMITRIJ *muffig* Ja, amüsier dich nur.
GOLDTBLOM *sendungsbewußt* Wir sind die Konservativen. Wir achten drauf, daß alles so bleibt, wie's ist.
Dimitrij schickt Goldtblom mit strengem Blick auf Streife durch die Ausstellung.
GOLDTBLOM *schreitet, bleibt stehen, guckt, schreitet usw., plötzlich* Didihhh. Die Fliegeruhr *(s. S. 198)* hat sich bewegt. *Liest ängstlich die Instrumentenanzeige ab* 48°54' nördliche Breite, 9°12' östliche Länge, 300 m über dem Meeresspiegel. Wir sind im Sinkflug.
DIMITRIJ *im Halbschlaf* Ein wunderbarerApparat. Achtet drauf, daß wir im nötigen Abstand zur Erde bleiben.
GOLDTBLOM *guckt weiter gebannt, erschrickt* Jetzt piepst er.

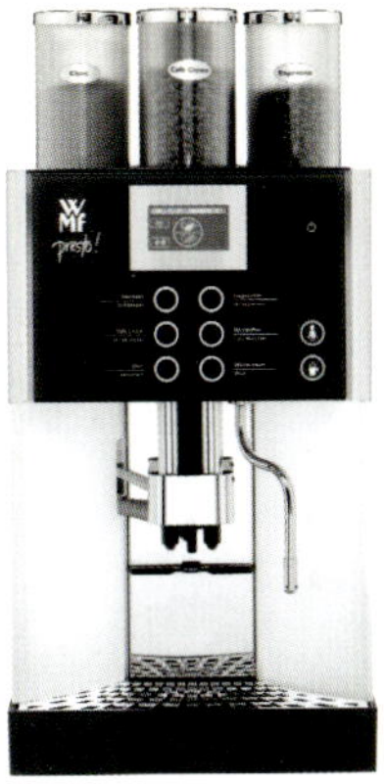

Characters: Dimitrij and Goldtblom (two aging attendants); Cleaning woman (dumb); Stewardess (boss of a party service, with a number of waitresses); Caretaker; Driver.

Stage setting: giant-sized automatic coffee machine ›WMF Presto‹ (see fig.) incorporating double glass doors at the front, positioned in such a way that the audience can see anyone coming; on stage stands the complete »Focus Energy 2006« exhibition.

Paying attention

—

Dimitrij and Goldtblom lying on the sofa (see page 134).

DIMITRIJ Listen when I tell you something.

GOLDTBLOM Sorry.

DIMITRIJ It's ok. I said there's nobody there.

GOLDTBLOM Where?

DIMITRIJ *seething* In the auditorium, you—*bleep*—. Where else?
And there's no one in the exhibition either.

GOLDTBLOM *unruffled* Supposing there was somebody there, what would we be doing then?

DIMITRIJ Watching.

GOLDTBLOM Watching who?

DIMITRIJ Watching out. If you want to get all clever on me then I'll start playing the boss, and you can start doing the rounds even if there is no one here.

Embarrassing silence

GOLDTBLOM Can you remember?

DIMITRIJ What? How we waited?

GOLDTBLOM No, how we froze.

DIMITRIJ Froze and waited.

GOLDTBLOM Well, we are not freezing anymore. It's warm here. Waiting is more pleasant, when there's heating.

DIMITRIJ But we are not waiting anymore. Now we have to pay attention.

GOLDTBLOM In this place, that is one and the same thing. We are attendants.

DIMITRIJ We have to make sure that none of the exhibits get damaged.

GOLDTBLOM *singing* »… Hush little baby, …«

DIMITRIJ *sulky* Go on, have your fun.

GOLDTBLOM *with an air of responsibility* We are the conservatives. We make sure that everything stays as it is.

With an admonishing glance, Dimitrij orders Goldtblom to do a round of the exhibition.

GOLDTBLOM *takes a walk, stops, looks, starts walking again, then suddenly* Didihhh! The aviators' instrument panel *(see page 198)* has moved. *Anxiously reads the instrument display* 48°54′ north, 9°12′ east, 300 m above sea level. We have begun the descent.

DIMITRIJ *half asleep* A fantastic instrument. Takes care that we keep off the ground.

GOLDTBLOM *continues to stare, gasps in a state of fright* It's beeping.

DIMITRIJ *dreht sich auf die Seite, gleichgültig* Fertigmachen zur Notlandung.
Die Raumpflegerin bugsiert ihre Gerätschaften zur Tür herein, grüßt wortlos.
Willkommen! Perfekt, die Bodentruppe ist prompt zur Stelle. *Goldtblom hin zum Eingang, hält die Tür auf; die Raumpflegerin mit ungelenken Gebärden.*

DIMITRIJ *übersetzt simultan aus dem Gestischen* Sehr aufmerksam, Herr Goldtblom. Dankeschön, Herr Goldtblom. Aber das wäre doch nicht nötig gewesen, Herr Goldtblom.

GOLDTBLOM Didi, hör auf. Jede Wette, die Raumpflegerin kennt die Ausstellung besser als wir zwei zusammen: jeden Winkel, den sie aussaugt; jedes Exponat, das sie abstaubt; jeden Sockel, um den sie herumwischt. Sie ist die Sorgfalt hoch drei. *Die letzten Worte gehen im Aufheulen des Staubsaugers unter. Goldtblom flüchtet sich zu Dimitrij aufs Sofa, lebhaftes Gespräch beider, vom Lärm übertönt. Das Staubsaugergeheul ebbt ab und die Raumpflegerin geht ab* … komm mir nicht mit so was. Aufmerksam sein ist kein leichtes Amt: immer einhergehen und den Blick streifen lassen und fokussieren wie ein Satelliten-Strahler *(s. S. 154)*, ständig mit den Augen herumwedeln, als wären sie leuchtdiodenbestückte Taschenlämpchen *(s. S. 160)*, alleweil auf dem Quivive, dauernd in Habacht. Ach, ich mag's nicht länger leiden. *Dimitrij ratlos. Pause* Mein Traum wär halt, ich verdiente mein Geld im Schlaf.

DIMITRIJ *unvermutet inspiriert* Guck mal, die ›Adam-Retorte‹ dort, der Holzkohlebrenner *(s. S. 56)*. Er ist so gebaut, dass die Verkohlung von Holz, einmal angeheizt, per Rauchgasverbrennung für sich selber sorgt. *Goldtblom wie vernagelt, Dimitrij bleibt geduldig* Oder, die Wasserspülung draußen auf dem Klo …

GOLDTBLOM Ja?

DIMITRIJ … hat ein Schwimmerventil eingebaut.

GOLDTBLOM Ja. Genau.

DIMITRIJ Wasserkasten voll – Schwimmer verschließt Zulauf.

GOLDTBLOM *erleuchtet* Wasserkasten leer – Schwimmer öffnet Ventil. Raffiniert. Fabelhaftes Design. Paßt auf sich selber auf. Die Aufmerksamkeit ist dem Gerät förmlich eingepflanzt.

DIMITRIJ Na endlich. – So. Und jetzt dasselbe für die Ausstellungsaufsicht.

GOLDTBLOM Nicht, daß wir uns hier selber wegautomatisieren.

DIMITRIJ Keine Bange. Wir liegen hier und stellen uns schlafend; die Besucher bemerken's und gehen fortan auf Samtpfoten und im Flüsterton durch den Exponatewald. Indem sie aufs eingeschlafene Wachpersonal Rücksicht nehmen, geben sie automatisch Acht auf die ausgestellten Dinge.

GOLDTBLOM Wir sind Zauberer.

DIMITRIJ Nein, Psychotechniker. Kybernetiker.

GOLDTBLOM Sag ich doch. Illusionisten.

DIMITRIJ *kichert in sich hinein* Die Besucher passen aufs Sofa auf. Und das Sofa wacht wieder über uns. Verkehrte Welt.

GOLDTBLOM Wir sind Revolutionäre – auf dem Kanapee. Ich träum wohl.
Bühnenbeleuchtung geht langsam aus, dabei Spieldosenmusik nach der Melodie »Weißt du, wieviel Sternlein stehen …«

DIMITRIJ *turns on his side, unconcerned* Prepare for an emergency landing.
The Cleaning lady manoeuvres her equipment through the door. Nods in greeting.
Hello! Perfect! The ground troops show up on time. *Goldtblom strides to the door, opens it; the Cleaning lady moves awkwardly.*

DIMITRIJ *translates her gestures into words* Oh, thank you, Mr. Goldtblom. Very attentive of you, Mr. Goldtblom. You didn't have to go to the trouble, Mr. Goldtblom.

GOLDTBLOM Didi, stop it. Bet you anything the Cleaning lady knows the exhibition better than the two of us put together: each corner she vacuums; each exhibit she dusts; each pedestal she mops around. She is thoroughness in person. *The last words are lost in the noise of the vacuum cleaner. Goldtblom flees to Dimitrij on the sofa, both get involved in a lively discussion, drowned out by the noise. The vacuum cleaner eases off and the Cleaning lady leaves …* don't come at me with that. Being an exhibition attendant is no easy job: always being on the prowl and keeping an eye on everything, focused like a reflector *(see page 154)*, shifting your attention from here to there like an LED torch *(see page 160)*, constantly on the alert, never letting your mind wander. Agh, I can't stand it much more. *Dimitrij puzzled. Pause* My dream was to earn money in my sleep.

DIMITRIJ *surprisingly inspired* Look at the ›Adam retort‹ there, the charcoal maker *(see page 56)*. It's made so that the wood turns to charcoal just from the heat generated from afterburning the exhaust. *Goldtblom makes no sign of movement, Dimitrij remains impatient* Or the toilet flush outside …

GOLDTBLOM Really?

DIMITRIJ … it's got a float built into it to close the valve.

GOLDTBLOM Yeah. Really.

DIMITRIJ Cistern full—float closes tap.

GOLDTBLOM *catches on* Cistern empty—float sinks, valve opens. Very clever. Fantastic design. Looks after itself. You could say that attentiveness is incorporated in the very design of the thing.

DIMITRIJ Now, you've got it.—So. And now the same for our job.

GOLDTBLOM Not that we want to automate our jobs away.

DIMITRIJ Don't worry. We'll lie here and pretend to sleep. The visitors will notice it and start tiptoeing their way through the exhibition, whispering among the exhibits. By being considerate to us, the sleeping attendants, they will automatically take care of the exhibits.

GOLDTBLOM We are magicians.

DIMITRIJ No, psychotechnicians, cyberneticians.

GOLDTBLOM Just what I said, conjurors.

DIMITRIJ *giggles to himself* The visitors keep an eye on the sofa. And the sofa keeps an eye on us. Crazy world.

GOLDTBLOM We are revolutionaries—man the settee! I must be dreaming.
Lights fade, music box plays the lullaby »Hush little baby …«

Politur

—

Dimitrij und Goldtblom zunächst beide auf dem Sofa (s. S. 134) liegend, später an der zweiflügeligen Glastür.

GOLDTBLOM *schreckt hoch* Entschuldigung. Wie spät?

DIMITRIJ *verschlafen* Es will Abend werden.

GOLDTBLOM *besorgt* Höchste Zeit, daß Häppchen und Trinkbarkeiten anrollen.

DIMITRIJ *gähnt ungeniert* Ich könnte jetzt auch eine Kleinigkeit vertragen.

GOLDTBLOM Blamier uns nicht. Die Sachen sind für die Gäste, Besucher, Preisträger. Zur Eröffnung nachher.

DIMITRIJ Welcher Öffnung?

GOLDTBLOM Er-öffnung. Hier. Von unserer Ausstellung. Hast du drauf vergessen?

DIMITRIJ Nein. Aber ich hab im Schlaf nicht mehr dran gedacht.

GOLDTBLOM Siehst du, ein Nickerchen macht alles neu.

DIMITRIJ Wir sehen unsere Ausstellung mit neuen Augen.
Wachsamkeit. Geistesgegenwart. Äußerste Sorgfalt. Alles wieder da. Alles noch da.

GOLDTBLOM Wir sind hochprofessionelle Wachleute.

DIMITRIJ Und wir sind gerüstet, wenn sie jetzt kommen.
Hinter der Glastür erscheinen die jungen Damen vom Partyservice.

GOLDTBLOM Wer kommt? *Die Aufwärterin rüttelt an der abgeschlossenen Tür.*

DIMITRIJ *ruft beflissen in Richtung Tür* Wir kommen. *Zu Goldtblom* Gogo, geh. Schließ auf.

GOLDTBLOM *geht unter unverständlichem Gemaule zur Tür* Na endlich.

AUFWÄRTERIN *dynamisch herein* Ah, die Herren Kollegen von der wachsamen Zunft.
Beim Frischmachen, wie ich sehe.

DIMITRIJ Schöpferische Pause.

AUFWÄRTERIN Dann könntet ihr, schön ausgeruht und erholt, uns beim Reintragen helfen?
Goldtblom will schon eilfertig nach draußen.

DIMITRIJ Eigentlich lieber nicht. Wir können sonst nicht aufpassen.

AUFWÄRTERIN *sprachlos, holt Luft* Und wenigstens die Tür aufhalten?

GOLDTBLOM *ganz Stimme seines Herrn* Bedaure, lieber nicht. Grüßen müßte genügen.
Das nämlich bewerkstelligen wir hier mit einem Winkkraftwerk *(s. S. 24)*.

AUFWÄRTERIN Jungs, ihr habt ein Rad ab.

GOLDTBLOM Ich wüßte nicht. Bei meinem letzten Rundgang war das Flügelrad noch dran.

DIMITRIJ *lenkt ein* Laß gut sein, Gogo. Wir könnten es weniger grobkörnig probieren. Geschliffener.
Und wenn wir nur so täten, als ob wir höflich wären.
Beide gehen zur Tür und halten je einen Flügel auf. Immer, wenn eine Bedienerin mit ihrem Servierwagen (s. S. 120) hereinrollt, verbeugen sie sich und grüßen.

GOLDTBLOM Nach dir, mit Verlaub.

DIMITRIJ *mit dem vorbildgebenden Mut des Vorgesetzten zur Blamage, bückt sich wenig und verquält*
Guten Tag, Frau Stellvertretende Directrice.

Polish

—

At first, Dimitrij and Goldtblom are both lying on the sofa *(see page 134)**, later standing at the glass doors.*

GOLDTBLOM	*jumps up suddenly* Sorry. What's the time?
DIMITRIJ	*drowsily* It must be evening.
GOLDTBLOM	*worried* High time to roll out the drinks and the finger food.
DIMITRIJ	*yawns unashamedly* I could do with a little something.
GOLDTBLOM	Don't embarrass us. The things are for the guests, the visitors, the prize-winners. For the opening.
DIMITRIJ	What opening?
GOLDTBLOM	The opening. Here, of our exhibition. Have you forgotten?
DIMITRIJ	No. But I wasn't thinking about it in my sleep.
GOLDTBLOM	See? A nap changes everything.
DIMITRIJ	We see our exhibition with new eyes. Attention. Presence of mind. Extreme care. It's all come back. It's all still there.
GOLDTBLOM	We are highly professional exhibition attendants.
DIMITRIJ	And we are equipped and ready should they arrive this minute.
	The young ladies from the party service appear behind the glass doors.
GOLDTBLOM	Who's there? *The Stewardess rattles the locked door.*
DIMITRIJ	*calls solicitously at the door* We're coming! *to Goldtblom* Gogo, go, unlock the door.
GOLDTBLOM	*goes to the door, grumbling loudly* Finally.
STEWARDESS	*walks in energetically* Ah, our dear colleagues, the professional watchdogs. Waking up, as I see.
DIMITRIJ	Just taking a short pause to find inspiration.
STEWARDESS	Now that you are all refreshed and rejuvenated, could you help us carry the things in?
	Goldtblom wants to rush outside.
DIMITRIJ	Actually, we'd rather not. Otherwise, we can't do our duty.
STEWARDESS	*speechless, draws a breath* At least hold the door open?
GOLDTBLOM	*in an officious tone* Very sorry, we'd rather not. A mere welcome will have to suffice. We do that here with a wind generator *(see page 24)*.
STEWARDESS	You guys are off your rocker.
GOLDTBLOM	Didn't realize that. During my last round I was rocking quite well.
DIMITRIJ	*gets involved* Leave it, Gogo. We could be a little less rough around the edges. More polished. Maybe if we even just pretend to be courteous.
	Both go to the doors and each holds one open. Each time a waitress comes in with a trolley (see page 120) they bow politely and welcome her officially.
GOLDTBLOM	Please, after you.
DIMITRIJ	*with the courage of a superior willing to make a fool of himself just to set an example for his subordinates, bows slightly and painfully* Have a nice day, honorable director.

AUFWÄRTERIN Wünsche gleichfalls einen schönen Tag, Herr Ausstellungsdirektorenanwärter.
DIMITRIJ *zu Goldtblom* Ich bring's nicht. Sie macht sich lustig.
GOLDTBLOM Wir machen sie fertig mit unserer Untertänigkeit.
DIMITRIJ *zitiert deklamierend* »Ich bin der erste Diener meines Stalls!« – Gestatten, Bauer Fritz, der mit den größten Kartoffeln.
GOLDTBLOM Also, auf zu fröhlicher Fron. Einfach tiefer beugen. Geschmeidiger. Galanter, wenn's geht. Vielleicht noch eine kleine einladende Handbewegung.
DIMITRIJ Mach mal erst selber.
GOLDTBLOM *zur Häppchenhereinbringerin, tiefer Diener mit ausschwingendem Arm* Einen schönen guten Abend, Frau Großgrundbesitzerswitwe.
DIMITRIJ Kein Mensch nimmt dir das ab. Das bist nicht du.
GOLDTBLOM Wenn ich ich sein wollte, wär ich Misanthrop geworden. *Zur Weinhereinfahrerin* Grüß Gott, Eilige Mutter. *Erntet ein huldvolles Lächeln*.
DIMITRIJ Schmiere, mein Lieber, reine Schmiere.
GOLDTBLOM Dann mimen wir es in der nächsten Runde mit weniger Öl.
DIMITRIJ *zur Tischdekorateurin* Willkommen, Gnädigste.
GOLDTBLOM Oha, wir finden ihn schon, den richtigen Ton, die rechte Dekoration. Wir sind Sozialdesigner.
DIMITRIJ ›Gehilfen‹ reicht. Wir halten halt die Wahrheit den Leuten so hin, daß sie bequem reinschlüpfen können.
GOLDTBLOM *altklug* Ganz nach Art jener Vase da *(s. S. 144)*; hört auf den sinnreichen Namen »Opening«; durfte, als sie noch jung und weich war, platzen; und streckt, gebrannt und gebändigt jetzt, dem Nutzer ihre gezierte Form entgegen. Vom einstigen Energieausbruch kündet ihr Ornament. Und das versöhnt den vom Purismus verprellten ›einfachen Mann aus dem Volke‹ mit ihrer Unterwerfung unter den Nutzen. Wohlverstanden, eine Geste der Höflichkeit.
DIMITRIJ Das reicht bequem für den Ethik-Ästhetik-Lehrstuhl an der Theaterakademie.
GOLDTBLOM *vor sich hin* Öl auf mein rostiges Gemüt. *Mitten in sein seliges Lächeln hinein Kurzschluß, totaler Lichtausfall*.
DIMITRIJ *unisono im Finstern, Theaterdonner im Hintergrund*
GOLDTBLOM Pause, meine Damen und Herren. Bleiben Sie dran. Wir sind gleich wieder für Sie da.

Genie

—

Geisterhafte Stimmung, undeutliche Geräusche aus der entfernten Eröffnungsfeier. An der Wand hinter dem Sofa eine hohe Bockleiter; ganz oben der Hausmeister; in halber Höhe Dimitrij, hält die Solarlaterne (s. S. 188), vorerst die einzige Lichtquelle.
GOLDTBLOM Sei gescheit, Didi, bleib unten. Wir sind so hoch nicht bezahlt.
DIMITRIJ Wir haben den Ausstellungsnotstand: Nichts zu sehen. Da ist jedes Mittel recht. Sogar guter Wille. *Hausmeister bringt Bohrmaschine zum Rasen*.

STEWARDESS I wish you a nice day too, honorable exhibition attendant supervisor.

DIMITRIJ *to Goldtblom* I can't do it. She's laughing at us.

GOLDTBLOM We will destroy her with our submissiveness.

DIMITRIJ *proclaims* »I am the top servant of my stall!« — May I introduce farmer Fred, the one with the biggest potatoes.

GOLDTBLOM Right, on to the delightful side of drudgery. Simply bow deeper. Smoother. More gallantly, wherever possible. Maybe a small welcoming gesture with the hand.

DIMITRIJ You first.

GOLDTBLOM *to a woman bringing in the finger food, deep bow with a flourish of the arm*
A wonderful evening to you, honorable dowager.

DIMITRIJ No one is going to believe you. That is just not you.

GOLDTBLOM If I had wanted to be me, I would have become a misanthropist.
To a woman delivering the wine A very good day to you, »Urgent Mary«.
Gets a gracious smile in return.

DIMITRIJ Greaser! Pure grease.

GOLDTBLOM Then we'll mime the next round with less oil.

DIMITRIJ *to the table decorator* Welcome, honorable lady.

GOLDTBLOM Oh yes, we'll get it. The right tone, the right flourish. We are social designers.

DIMITRIJ ›Assist‹ is enough. We'll simply hold up reality in a way that the people can slip into it comfortably.

GOLDTBLOM *precocious* Just like that vase there *(see page 144)*; bears the wonderfully rich name of »Opening«; when young and tender it was allowed to burst forth; and now, fired and tamed, it stretches its decorated form towards the user. Its ornament speaks of past explosions of power. And yet its submission to everyday use soothes the sensibilities of the common man, bruised as he is by the barrage of minimalist design. A gesture of courtesy, let there be no doubt.

DIMITRIJ Hey, that'll easily get you the chair for ethics and aesthetics at drama school.

GOLDTBLOM *to himself* Oil to my rusty steed.
In the middle of his satisfied smile, a short circuit plunges the stage into darkness.

DIMITRIJ *uni sono in the darkness, rumblings heard in the back of the theatre*

GOLDTBLOM A short break, Ladies and Gentlemen. Stay where you are. We will be back in a moment.

Genius

—

Ghostly atmosphere, muffled sounds of a distant opening ceremony. On the wall behind the sofa a high stepladder; Caretaker is positioned at the top; Dimitrij is halfway up, is holding a solar-powered lamp (see page 188), at first the only source of light.

GOLDTBLOM Don't be stupid, Didi, stay down. We are not paid that much money.

DIMITRIJ We have an exhibition state-of-emergency: can't see a thing. Whatever works will do. Even if it's just goodwill. *The Caretaker starts up the drill.*

GOLDTBLOM Halt. Stop. Bohrmehl auf dem Sofa. Die Firma möcht heikel sein.
Reicht den Bohrstaubsauger (s. S. 42) nach oben. Hausmeister wendet an.

DIMITRIJ Ah, welche Verfeinerung des Bohrens.

GOLDTBLOM Jetzt noch die Wiha-Zange, und der minimalinvasive Eingriff in die Wand ist perfekt.
Reicht dem Hausmeister die Kombizange (s. S. 46).

HAUSMEISTER Lacht nur, Narren. Was ich in der Hand halte, ist Intelligenzverstärkung pur. *Doziert von der Leiter herab* Hier *hält seine Hand wie Albrecht Dürer die seine auf der Zeichnung von 1493* feinfühligster Präzisionsgriff! Und was die Evolution hier drin ersonnen hat, hat der Werkzeugmacher in den Zangenkopf hinaus projiziert – Hier *faßt die Bohrmaschine in Arbeitshaltung* energischer Kraftgriff! Von dieser Erfindung der Natur haben die Ingenieure die Zangengriffstücke abgegossen – und beide Abformungen dann im neuen Werkzeug zusammentreffen lassen. Noch so ein raffiniertes Kräfteparallelogramm dazwischen eingefügt, und …

GOLDTBLOM *platzt vor Begeisterung* … und fertig war die Blitzgescheitheit. Dieselbe Genialität, Didi, mit der sie dich und mich zum Aufpassen in die Ausstellung hier zusammengesteckt haben.

HAUSMEISTER *werkelt weiter am Bohrloch* So. Saft wäre wieder da. *Ausstellungsbeleuchtung flackert auf. Hausmeister von der Leiter, packt sein Werkzeug zusammen.*

GOLDTBLOM *zu Dimitrij* Hast du gehört? Oben in der Eröffnungsfeier, der hinreißend fürsorgliche, äußerst zuvorkommende und wunderbar kluge Schlußsatz?

DIMITRIJ Hm?

GOLDTBLOM »Das Buffet ist eröffnet.«

DIMITRIJ *hat die Scherze satt* Geh, Gogo. Aufschließen. Beide Flügel weit auf.
Andrang der Gäste an Buffet, Exponaten, Verkaufstisch. Aufsichtskräfte geben nach Kräften Auskunft, verkaufen Kataloge, weisen unbotmäßige Besucher zurecht usw. Hausmeister mit Leiter strebt vergeblich zum Ausgang. Das Ganze ein lärmiges Ballett. Mitten im Trubel brettert die Feuerwehr (s. S. 180) auf die Bühne, hält mit quietschenden Reifen vor dem Sofa.

FAHRER *zum Fenster heraus* Tag der offenen Tür! Probefahrt!
Alle Anwesenden drängeln in das Fahrzeug, Dimitrij und Goldtblom vorneweg. Feuerwehr ab wie die Feuerwehr, gibt Blick wieder frei aufs Sofa (s. S. 134). Wie hingezaubert stehen die Aufsichtskräfte drauf, hüpfen, wie es eben gesetzte Herren vermögen, allmählich immer höher.

GOLDTBLOM Wir brauchen nicht mehr aufzupassen …

DIMITRIJ … gegen niemanden mehr höflich sein …

GOLDTBLOM … wie nicht recht gescheit.
Mit dem letzten energischen Sprung entschwinden sie in den Theaterhimmel.

GOLDTBLOM Stop! You're getting dust on the sofa. The company would like to maintain its standards. *Passes up the drilling dust collector (see page 42). Caretaker uses it.*

DIMITRIJ Ah, what an improvement to the art of drilling.

GOLDTBLOM Now, the Wiha Professional, and the minimally invasive operation on the wall is accomplished. *Hands the combination pliers to the Caretaker (see page 46).*

CARETAKER Go on, laugh, you fools. Whatever I hold in my hand is a perfect extension of my intelligence. *Lectures the others from the ladder* Here, *holds his hands like in Dürer's sketch from 1493.* Finest precision touch! And what evolution was thinking about here is what the toolmaker projected into the head of the pliers. Here, *grabs the drill as though he were about to use it* what a powerful grip! Based on this invention of Mother Nature, the engineers have formed the handles—and both these shapes are combined perfectly in the new tool. Another refined parallelogram of forces fitted in between, and …

GOLDTBLOM *bursts out with delight* … and there it is, a stroke of genius! The same brilliance, Didi, that brought us together to be the attendants of this exhibition.

CARETAKER *continues working on the hole* There we are. The power is back on.
Exhibition lighting flickers back on. Caretaker gets down from the ladder, packs his tools up.

GOLDTBLOM *to Dimitrij* Did you hear that? Upstairs in the opening ceremony, that exquisitely caring, beautifully considerate and wonderful closing line?

DIMITRIJ Hm?

GOLDTBLOM »The buffet is open.«

DIMITRIJ *sick of the joke* Go, Gogo, go, unlock the doors, open them wide.
Crush of guests at the buffet, the exhibits, promotional table. The attendants do their best to provide information, sell catalogs, warn unruly visitors to get in line, etc. Caretaker, carrying the ladder, struggles vainly to get to the door. The whole scene resembles a chaotic ballet. In the middle of it all a fire engine storms on stage (see page 180), tires squealing, comes to a halt in front of the sofa.

DRIVER *calling out the window* Open day! Who wants a ride?
All present struggle to get in the vehicle, Dimitrij and Goldtblom up the front. Fire engine leaves as it came, revealing the sofa once more (see page 134). As if in a trance, the attendants get up on it, start jumping, as respectable gentlemen are wont to do, slowly getting higher and higher.

GOLDTBLOM We don't have to attend to anyone any more …

DIMITRIJ … no longer need to be polite to anyone …

GOLDTBLOM … like lunatics.
And with their last jump they are lifted up into the ceiling.

Focus Energy
Der Wettbewerb

Sabine Lenk
Projektleitung Focus Energy
Design Center Stuttgart
—

Steigende Rohstoff- und Energiepreise, das Ende des Ölzeitalters, das Wissenschaftler proklamieren, politische Konflikte um Energiequellen, die Suche nach neuen Energieträgern, die Förderung und Erforschung erneuerbarer Energietechnologien – das sind nur einige der Aspekte, die deutlich machen, dass die Energiefrage eine der wichtigsten und vordringlichsten Zukunftsfragen ist, die gelöst werden muss.

So lag es nahe, dieses Jahr den Wettbewerb für den Internationalen Designpreis Baden-Württemberg unter das Thema »Focus Energy« zu stellen. Denn auch nahezu alle Produkte aus den verschiedensten Branchen müssen sich verstärkt der Energiefrage stellen. Im Markt werden immer mehr energiesparende Produkte nachgefragt und solche, auf deren Verbrauch der Nutzer Einfluss nehmen kann.

Materialeinsatz und Energiebedarf in der Herstellung spielen unter ökologischer und ökonomischer Sicht zunehmend eine entscheidende Rolle. Dies gewinnt für Designer und Unternehmen zukünftig noch stärker an Bedeutung. Die ausgezeichneten Produkte belegen denn auch, dass zukunftsweisende, herausragende Produktgestaltung in engem Kontext zu Energieverbrauch und Energienutzung steht.

Der Internationale Designpreis Baden-Württemberg wird jährlich als Staatspreis für zukunftsweisende und herausragende Gestaltungsleistungen an Hersteller und Designer verliehen. Der renommierte Wettbewerb blickt auf eine lange Tradition zurück und genießt im In- und Ausland hohes Ansehen. Seit geraumer Zeit wird er unter einem Thema ausgeschrieben, das jährlich wechselt – eine besondere Herausforderung für die Einreicher. In der Vergangenheit waren das beispielsweise Themen wie Arbeitswelten, Mobilität, Lebensart, Balance, Dialog oder Know-how.

Gut gestaltete Produkte erhalten den »Focus in Silber«, die besten davon – pro Kategorie jedoch nur ein Produkt – den »Focus in Gold«. Die beiden Auszeichnungen sind begehrte Bestätigungen für höchste Designqualität. Für Unternehmen und Designer prämierter Produkte ergeben sich daraus klare Vorteile im Markt, denn der »Focus in Gold« und der »Focus in Silber« bürgen für höchste Produktqualität und unterstützen wirkungsvoll die Marktkommunikation. Die nichtkommerzielle Ausrichtung des Wettbewerbs ermöglicht auch kleinen und mittleren Unternehmen die Teilnahme.

Seit 2002 wird auch der Mia Seeger Preis für Nachwuchsdesigner unter dem gleichen thematischen Dach ausgeschrieben. Hiervon profitieren alle Beteiligten. Die Junior-Designer zeigen interessante konzeptionelle Ideen und Visionen, die Profis des Internationalen Designpreises präsentieren Serienprodukte mit Weltklasseniveau – eine gelungene Ergänzung.

Beim Internationalen Designpreis Baden-Württemberg 2006 konnten Firmen und Designer aus aller Welt für den Wettbewerb »Focus Energy« industrielle Serienprodukte einreichen, die nicht länger als zwei Jahre auf dem Markt sind. Zugelassen waren auch Prototypen, deren Serienreife gewährleistet war, jedoch keine Unikate und keine künstlerischen Arbeiten.

Focus Energy
The competition

Sabine Lenk
Project management Focus Energy
Design Center Stuttgart
–

Soaring raw materials and energy prices, scientists proclaiming the end of the oil age, political conflicts over access to energy sources, the search for new sources of energy, the promotion of and research into technologies to exploit renewable energy—these are just some topics that clearly illustrate that the energy issue is one of the most important and urgent matters we will have to solve in the future.

»Focus Energy« was therefore an obvious choice as the topic for this year's Baden-Württemberg International Design Award. After all, nearly every product, from the most diverse industries, is increasingly having to face up to the energy issue. On the market, demand for energy-saving products is growing all the time, as well as for products whose energy consumption can be influenced by the user.

Use of materials and energy consumption in production are increasingly playing a decisive role, both ecologically and economically. This is a subject that will become even more significant for designers and companies in the future. This year's award-winning products thus document how ground-breaking, excellent product design is closely related to energy consumption and energy use.

The Baden-Württemberg International Design Award is given annually to manufacturers and designers as a government prize for forward-looking and outstanding design work. This renowned competition can look back on a long tradition and enjoys an excellent reputation both in Germany and abroad. For some time now, the competition has been centred around a theme which changes from year to year—quite a challenge for entrants. In the past, these themes have included working environments, mobility, way of life, balance, dialog or know-how.

Well designed products are awarded the »Focus in Silver«, while the best products—but only one product per category—receive the »Focus in Gold«. Both awards are highly sought after as confirmation of excellent design quality. The awards given to products are a clear competitive advantage for companies and designers, for the »Focus in Gold« and the »Focus in Silver« are a guarantee of the highest product quality, and are an effective aid to marketing communication. The competition's non-commercial design allows small and medium-sized enterprises to participate.

Since 2002, the Mia Seeger Prize for young design talent has invited entrants to compete under the same theme. All those involved profit from this. The young designers present interesting conceptual ideas and visions, while the professionals entering for the International Design Award present world-class series-manufactured products—and the two sides complement each other perfectly.

For the 2006 Baden-Württemberg International Design Award, companies and designers from around the world were allowed to enter industrial series-manufactured products for the »Focus Energy« competition provided the products had not been on the market for longer than two years. While prototypes whose readiness for series production was guaranteed could also be entered, one-off products and artistic studies could not.

Eine weitere Bedingung war, dass Produkte der Juroren nicht am Wettbewerb teilnehmen konnten; denn beim Internationalen Designpreis Baden-Württemberg wird größter Wert auf Neutralität und Objektivität der Juroren gelegt. Außerdem sollten die eingereichten Produkte sich in eine der folgenden 12 Kategorien einordnen:

- Produktion, Montage, Logistik
  Maschinen, Werkzeuge, Vorrichtungen…
- Medizin, Rehabilitation, Geriatrie
  Diagnose- und Analysegeräte, medizinische Apparaturen…
- Haustechnik und Steuerung
  Heizungen, Klimatechnik, mechanische und elektronische Steuerungen…
- Bad, Sanitär
  Badeinrichtungen und -produkte…
- Küche, Haushalt
  Kücheneinrichtungen, Haushaltsgeräte…
- Interior
  Möbel, Teppiche, Objektausstattungen…
- Ambiente
  Büro-, Küchen-, Haushaltsprodukte, Geschirr, Interior Accessoires, Dekorationen…
- Beleuchtung
  Lampen, Leuchten…
- Kommunikation, Optik
  Informations- und Kommunikationsprodukte, Media-Produkte, Kameras…
- Architektur, öffentlicher Raum, Ausstellungen
  Gebäudekonzeptionen, Messebau, Orientierungssysteme…
- Transport, Verkehr
  Individualfahrzeuge, öffentliche Verkehrsmittel…
- Freizeit, Outdoor
  Sport- und Wellnessprodukte, Taschen, Spielzeug, Gartenartikel…

Insgesamt wurden 326 Produkte von Einreichern aus der ganzen Welt angemeldet. Dabei stammen die ausländischen Hersteller und Designer dieses Jahr aus Italien, der Schweiz, Österreich, aus den Niederlanden, Schweden, Belgien, England, Dänemark, Japan und Taiwan.

Zwei Tage begutachteten und prüften sechs international erfahrene und anerkannte Juroren aus unterschiedlichen Designdisziplinen die 326 Produkte im Original. Dabei gingen sie nach folgenden Kriterien vor:

- Gestaltungsqualität
- Innovationsgehalt
- Ergonomie
- Gebrauchsqualität
- Benutzerführung, Produktgrafik beziehungsweise -typographie
- Verwendung neuer Werkstoffe
- ökologischer Aspekt
- gestalterische Qualität der Produktausstattung: Bedienungs-, Montageanleitung, Verkaufsverpackung
- Wertigkeit, Anmutung und emotionaler Gehalt

Aus den eingereichten Produkten wählte die Jury 84 aus, die sie für auszeichnungswürdig hielt:

9 erhielten den »Focus in Gold« und 75 den »Focus in Silber«. Alle 84 Produkte sowie die Wettbewerbsergebnisse des Mia Seeger Preises sind im Jahrbuch publiziert und vom 21. Oktober bis zum 10. Dezember 2006 im Werkzentrum Weststadt in der Film- und Medienstadt Ludwigsburg ausgestellt.

One further condition was that the judges' own products were barred from competing; the Baden-Württemberg International Design Award places great importance in its judges' neutrality and objectivity. In addition, the products submitted had to fit into one of the following 12 categories:

— Production, assembly, distribution
machinery, tools, devices…
— Medicine, rehabilitation, geriatrics
diagnostic appliances and analysis devices, medical apparatus…
— Installations technology and controls
heating systems, air-conditioning, mechanical and electronic controls…
— Bathrooms, sanitary installations
bathroom fittings and products…
— Kitchen, household
kitchen furniture, household appliances…
— Interiors
furniture, carpets, contract furniture…
— Ambience
office, kitchen and household products, chinaware, interior accessories, decorations…
— Lighting
lamps, luminaires…
— Communication, optical equipment
information and communication products, media products, cameras…
— Architecture, public space, exhibitions
building design, trade fair structures, signposting systems…
— Transport, traffic
private and public transport…
— Leisure, outdoor
sport and wellness products, bags, toys, garden articles…

Entrants from all over the world submitted a total of 326 products. This year, the non-German manufacturers and designers participating in the competition come from Italy, Switzerland, Austria, the Netherlands, Sweden, Belgium, the UK, Denmark, Japan and Taiwan.

For two days, six internationally experienced and recognized judges from a range of design disciplines appraised and assessed the 326 products in the original. Their assessment was based on the following criteria:

— design quality
— innovative content
— ergonomics
— quality of use
— user controls, product graphics and typography
— use of new materials
— ecological aspect
— design quality of product peripherals: instructions for use and assembly, sales packaging
— quality, grace and emotional content

From the products submitted, the judges chose 84 that they felt were worthy of commendation:

9 received the »Focus in Gold« and 75 the »Focus in Silver«. All 84 products and the results of the competition for the Mia Seeger Prize are published in the yearbook, and will be on show in the Werkzentrum West in Ludwigsburg, a town renowned in the film and media world, from 21 October to 10 December 2006.

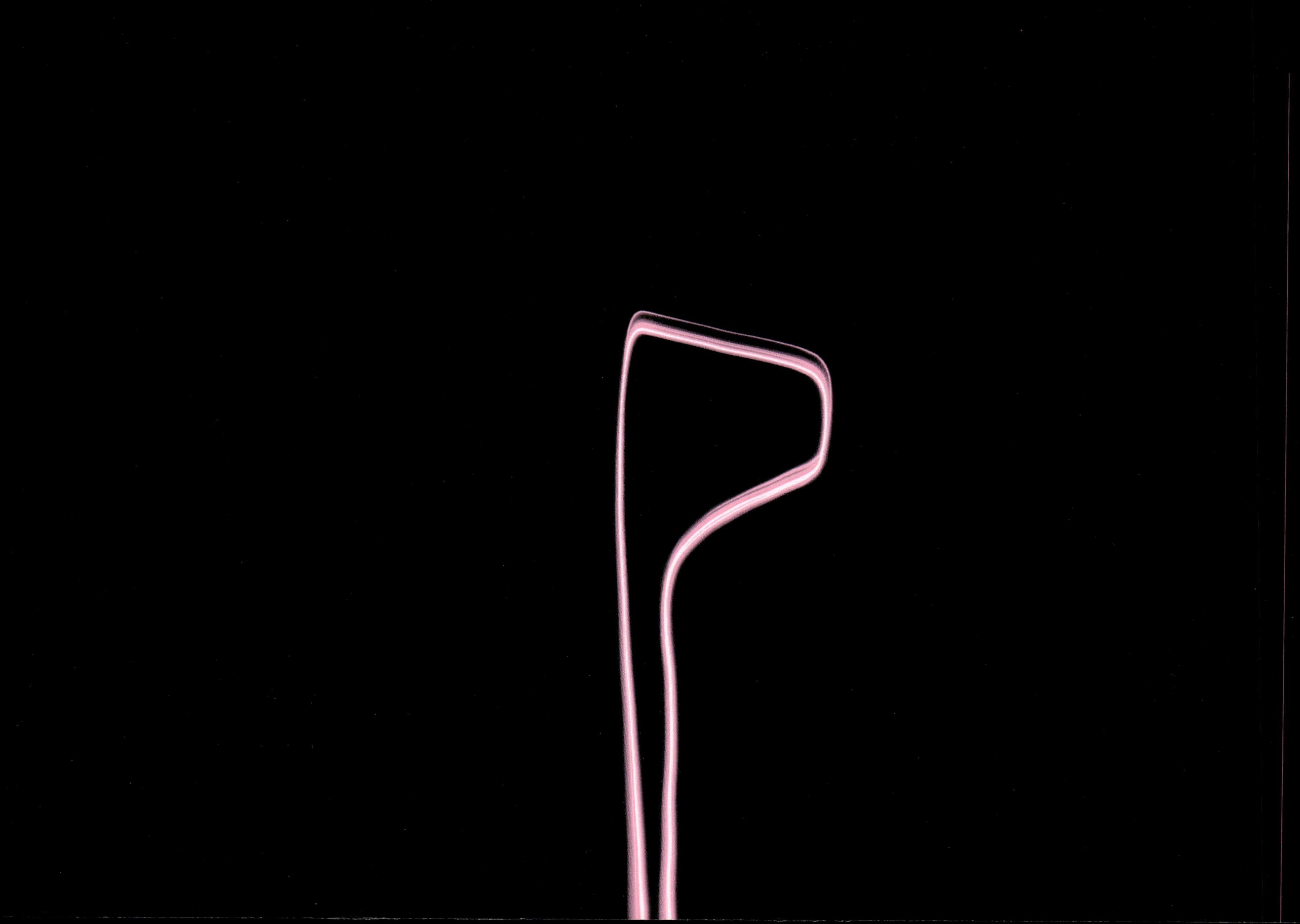

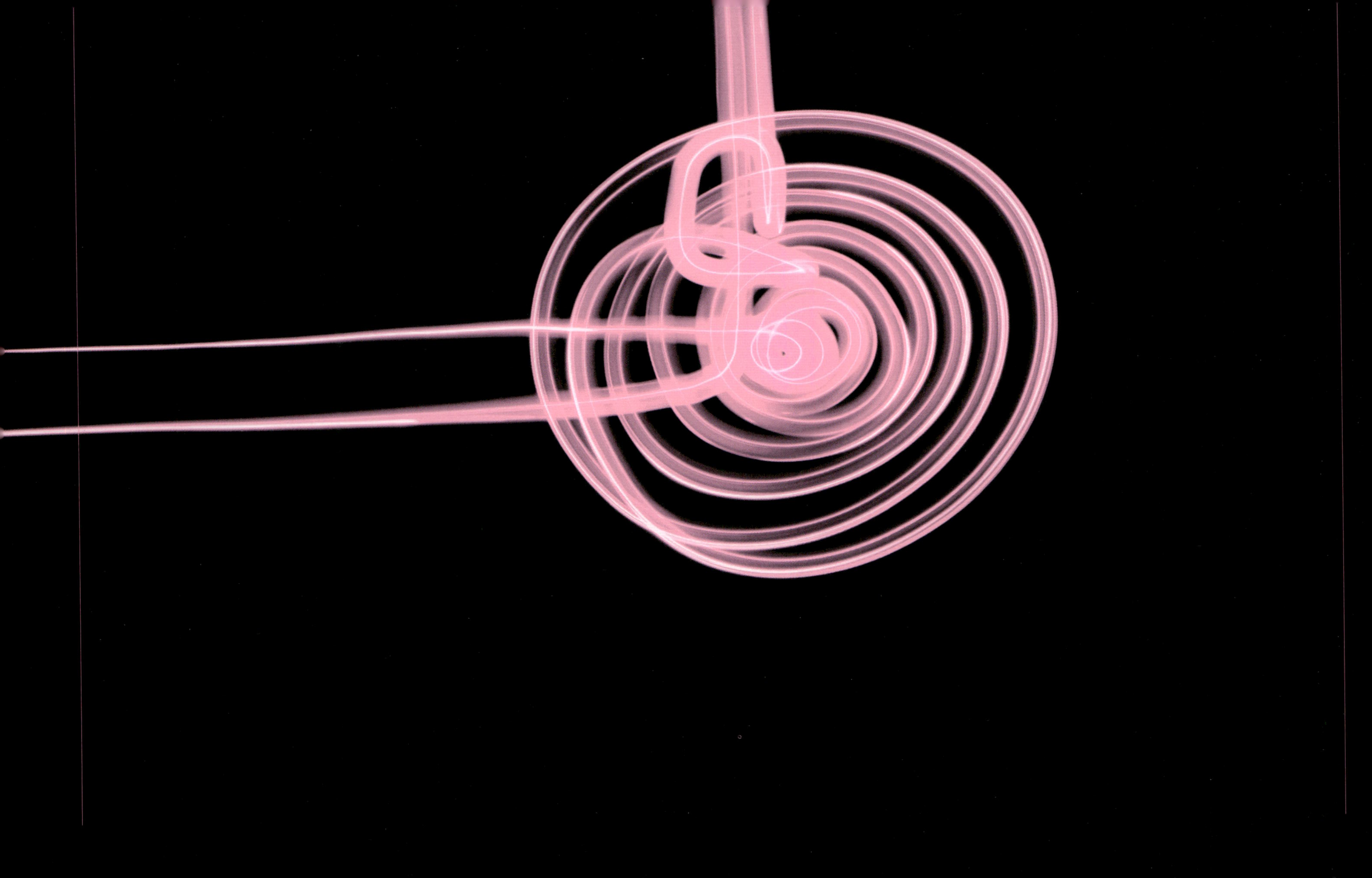

Conergy SWT 6000 AC
Kleine Windkraftanlage / Small wind turbine

Haustechnik und Steuerung / Gold
Installations technology and controls / Gold

CONERGY

Conergy SWT 6000 AC

Kleine Windkraftanlage
Small wind turbine

Hersteller / Manufacturer
Conergy AG
D-20537 Hamburg
—
Design / Designer
Werksdesign / In-house design
Sven Dähne, Peer Stock, Martin Dorsch
—
Vertrieb / Distributor
Conergy AG
D-20537 Hamburg
—

Die Windkraftanlage mit einer Leistung von 6 beziehungsweise 7,5 Kilowatt bietet Menschen Zugang zu Elektrizität, denn sie ist für autarke Stromversorgung in netzfernen Gebieten konzipiert: für Landwirte, kleine Unternehmen, netzferne Häuser oder in südlichen Ländern, beispielsweise Entwicklungsländern, für Dorfelektrifizierung, industrielle Anwendungen oder Telekommunikation.

Zudem kann sie aber auch für Netzanschluss und -einspeisung eingesetzt werden. Und sie schont gleichzeitig Ressourcen und Umwelt. Die Leichtbauweise unterstützt diesen Aspekt zusätzlich, reduziert Kosten und vereinfacht nachhaltig Herstellung, Transport, Wartung und Montage. Dank der Seilwinde kann die Anlage einfach und schnell von nur drei Personen an einem Tag sowohl auf einem Beton- als auch auf einem Stahlfundament montiert werden.

Im Gegensatz zu herkömmlichen Windkraftanlagen übersteht die Anlage auch stärkste Stürme. So dreht die patentierte passive Pitch-Regelung bei hohen Windgeschwindigkeiten die Rotorblätter aus dem Wind und gewährleistet die Stabilität der Anlage bis zu einer Windstärke von 250 Stundenkilometern. Das Ergebnis ist eine Anlage mit großer Widerstandsfähigkeit und Zuverlässigkeit mit wenig Materialeinsatz. Die Windkraftanlage bringt einen hohen Ertrag, sie produziert je nach Ort durchschnittlich 5000 bis 12000 Kilowattstunden pro Jahr.

This wind turbine, with a power output of 6 or 7.5 kW, is intended as a stand-alone generator to provide electricity in remote areas—farmers, small businesses, houses a long way from the grid—or, for instance, in developing countries, for village electrification, industrial applications or telecommunications.

However, it can also be connected to the grid and feed power into it. It is also eco-friendly, helping to limit depletion of resources. The lightweight design is a further step in the same direction, reducing costs and permanently facilitating production, transport, servicing and installation. Thanks to the winch, the turbine can be easily and quickly set up by three people in just one day, on either a concrete or a steel foundation.

Unlike conventional wind turbines, this unit withstands even the most violent storms. In high winds, the patented passive pitch control turns the rotor blades out of the wind, thus ensuring that the turbine remains stable up to a wind velocity of 250 km/h. The result is a very robust and reliable system built with low consumption of materials. The wind turbine is a high-yield system: depending on the location, it produces between 5 000 and 12 000 kWh of electricity a year.

Jury
Neuartiges Windrad, macht Windenergie für Regionen, die fern vom Netz sind, verfügbar. Erfordert weder große Flächen noch besonders schwierige Zulassungsverfahren, da es kleiner und leichter ist. Interessant auch, dass es schafft, sich dem stürmischen Wind durch Drehen der Rotorblätter anzupassen. Damit kann es auch noch bei hohen Windgeschwindigkeiten Energie erzeugen.
—

Judges panel
Novel wind turbine, makes wind energy available in regions remote from the grid. Is not space-hungry, nor are any complex approval procedures necessary, as it is relatively small and light. Also interesting that it can adapt to storm winds by turning its rotor blades, so that it can also generate energy at high wind velocities.
—

Jury
Adäquates Design für die Montage in Garagen und Nutzräumen, staubfest, schlagzäh, modular addierbar, gutes Design.

—

Judges panel
Appropriately designed for installation in garages and utility rooms, dustproof, knock-resistant, modular, good design.

—

Power Box

Solarbetriebener Stromversorger
Solar-powered power pack

Hersteller / Manufacturer
Würth Solergy
D-71672 Marbach am Neckar
–
Design / Designer
KPG Design Group GmbH
D-73527 Täferrot
–
Vertrieb / Distributor
Würth Solergy
D-71672 Marbach am Neckar
–

Mit der stationär montierten Stromversorgungseinheit können kleinere bis mittlere Stromverbraucher in Umgebungen ohne Strom betrieben werden, zum Beispiel Garagentorantriebe, Alarmanlagen, Belüftung, Funk- und TV-Geräte. Aufgeladen wird sie über ein Solarmodul.

Das Besondere ist, dass die Sonnenenergie, die durch ein Photovoltaik-Modul in Elektrizität gewandelt wird, in einem abnehmbaren, aus zwei Teilen bestehenden Akkumulatoren-System gespeichert wird, das auch am Netz wieder aufladbar ist. Das garantiert eine permanente, tageszeit- und witterungsunabhängige Versorgung.

Alle Prozesse, von der Umwandlung des Sonnenlichts in Energie, bis hin zur Abgabe an den Verbraucher, werden elektronisch gesteuert, was den Wirkungsgrad erheblich verbessert und für Langlebigkeit und Bedienerfreundlichkeit sorgt. Die »Power Box« besteht außen aus zwei Akku-Einheiten mit visueller und akustischer Zustandsanzeige und jeweils zwei Ladebuchsen. In der Mitte befindet sich eine Steuer/Ladeelektronik-Einheit mit visueller und akustischer Warnmeldung.

This power pack, which is installed as a stationary unit, allows small to medium-sized power consumers, such as garage door drives, alarm systems, ventilation systems, radio or TV sets, to be operated where there is no power supply. It is charged by means of a solar module.

What is special about this unit is that the solar energy, which is converted into electricity by a photovoltaic module, is stored in a removable, two-part rechargeable battery system, which can also be recharged from the mains. This guarantees a permanent power supply, independent of time of day and the weather.

All processes, from the conversion of sunlight into energy to the supply to the consumer, are controlled electronically. This considerably improves its efficiency, and means both a long service life and user friendliness. The »Power Box« comprises three sections. The two outside sections are the rechargeable batteries, whose charge status is signalled visually and acoustically, and which each have charging sockets. The middle section is an electronic control/charge unit with a visual and acoustic alarm function.

Jury

Macht die Solarpanels auf dem Dach akzeptabler für eine größere Zielgruppe, weil dies hier eine sehr flache Lösung ist, die sich auch vom Design her an die Dachpfannen anpasst, auch farblich. Insofern eine sehr gelungene ästhetische Lösung für diese regenerative Energie.

—

Judges panel

Makes solar panels on the roof more acceptable for a larger target group, since this is a very flat solution whose design also harmonizes well with roof tiles, also in terms of colour. In this respect, a fine example of an aesthetic solution for this renewable energy source.

—

Vitosol 100
Flachkollektor / Flat-plate collector

Vitosol 100

Flachkollektor
Flat-plate collector

Hersteller / Manufacturer
Viessmann Werke GmbH & Co. KG
D-35107 Allendorf/Eder
—
Design / Designer
Phoenix Design
D-70376 Stuttgart
—

Da die Dachgestaltung innerhalb der Gesamtarchitektur eines Hauses von großer Bedeutung ist, wurde der Flachkollektor vor allem unter Berücksichtigung architektonischer Gesichtspunkte entworfen. Architektonische Klarheit, feine Rahmengestaltung und sich sanft zu den Dachziegelreihen hin neigende Seitenflanken sind die charakteristischen Merkmale von »Vitosol 100«, bei dem die formale Integration in die bestehende Dachstruktur im Vordergrund stand. Das wird noch unterstrichen durch die Farbgestaltung, die sich an die jeweilige Dachgestaltung anpasst; denn auf Wunsch ist der Rahmen in allen RAL-Farbtönen lieferbar. Die hochselektive Sol-Titan-Beschichtung nutzt die Sonneneinstrahlung effizient aus und erzielt einen hohen Wirkungsgrad.

»Vitosol 100« ist besonders montagefreundlich. Das geringe Gewicht des Kollektors von 45 Kilogramm erleichtert den Transport auf das Dach und die Montage. Edelstahl-Wellrohr-Steckverbinder sorgen für einfache und schnelle Montage der Kollektoren.

As roof design is extremely important for the overall architectural appearance of a house, special attention was given to architectural considerations when designing this flat-plate collector. The characteristic features of »Vitosol 100«—architectural clarity, a light framework structure and sides that slope gently to the roof tiles—are evidence of the importance attached to formally integrating the collector with the existing roof structure. This is underscored further by the colour scheme, which can be adapted to any roof design, since the frame can be supplied in any of the RAL colours on request. The highly selective Sol-titanium coating means that the panel's exploitation of solar irradiation is highly efficient.

»Vitosol 100« is especially simple to install. As the collector weighs just 45 kilograms, carrying it up to the roof and installing it is easy. The stainless steel connector pipes allow individual collectors to be linked together quickly and simply.

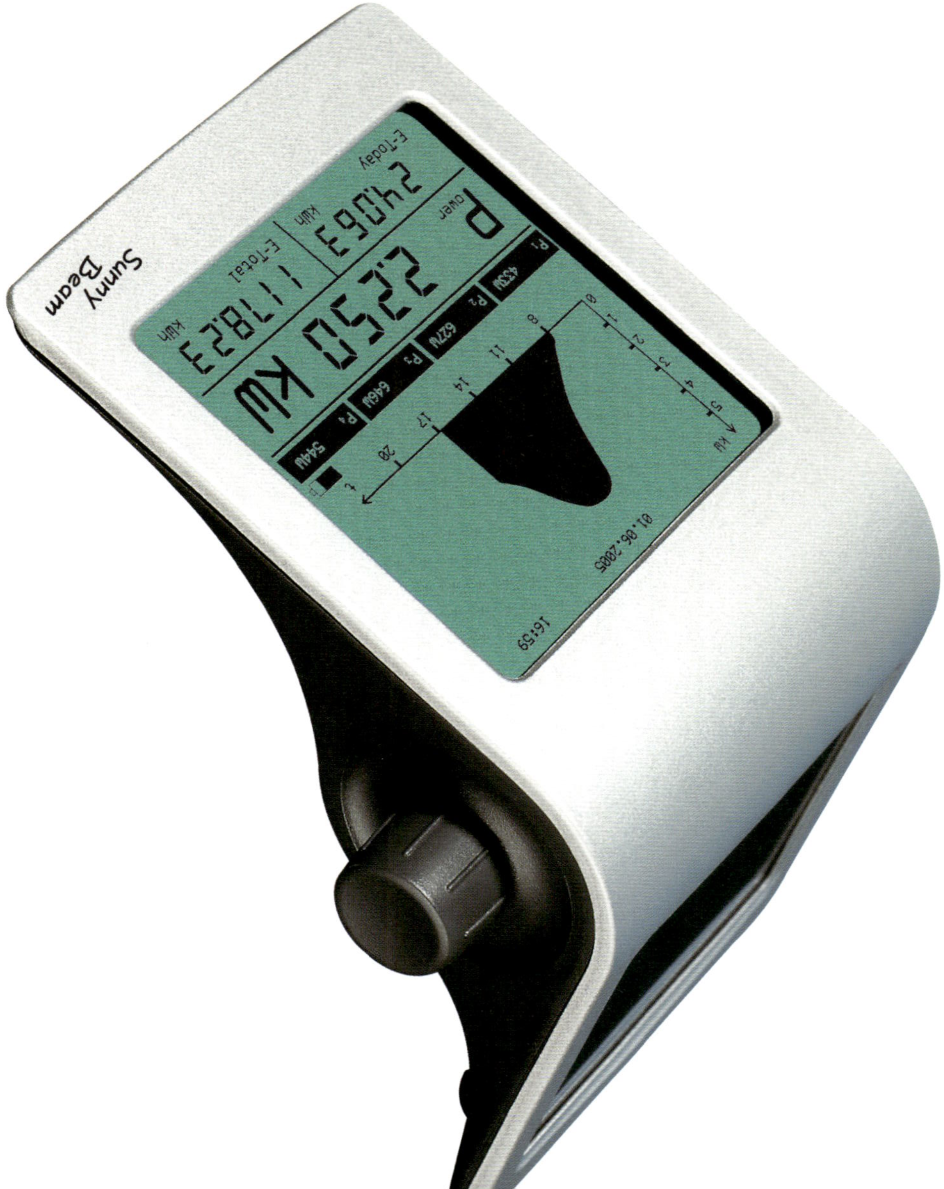
Sunny Beam
01.06.2005
16:59
P1 433W
P2 627W
P3 646W
P4 544W
Power
2.250 kW
E-Today
24.063 kWh
E-Total
1178.23 kWh

Sunny Beam

Solaranlagen-Überwachungs- und Steuerungsgerät

Monitoring and control unit for photovoltaic systems

Hersteller / Manufacturer
SMA Technologie AG
D-34266 Niestetal
—
Design / Designer
Modus Product Design GmbH
D-70435 Stuttgart
—
Vertrieb / Distributor
SMA Technologie AG
D-34266 Niestetal
—

»Sunny Beam« überwacht und steuert drahtlos Solaranlagen. Die Kommunikation mit den Wechselrichtern geht über Funk. Mit dem Gerät kann man zentral und zu jeder Zeit komfortabel den Status und den Ertrag der Wechselrichter kontrollieren. Es zeigt unter anderem den Tagesverlauf als Grafik, die aktuelle Leistung, den Gesamtenergieertrag, den laufenden Tagesenergieertrag sowie den Monatsverlauf in Form eines Balkendiagramms oder die eingesparten CO_2 Emissionen.

Das kompakte und leichte Gerät mit einem großen, leicht ablesbaren LCD-Bildschirm ist einfach und intuitiv mit dem Dreh-Drück-Knopf zu bedienen. Dank seines attraktiven, zeitlosen Designs kann »Sunny Beam« überall im Haus platziert werden, als Tischgerät oder mit einer Halterung an der Wand. Das Gerät wird über eine integrierte Solarzelle und einen Akku für Tage mit weniger Sonneneinstrahlung mit Energie versorgt. Über eine USB-Schnittstelle können Daten auf den PC übertragen und der Akku geladen werden.

»Sunny Beam« provides wireless monitoring and control of photovoltaic systems. Radio is used to communicate with the inverters. The status and yield of the inverters can now be checked conveniently at any time and by just one unit. The unit logs data such as the record for the day in graph format, current output, total energy yield, energy yield for the current day or for the current month in bar-chart format, or the CO_2 emissions saved.

With its large, easy-to-read LCD monitor, this compact, light unit can be operated simply and intuitively by means of a selector knob. Its attractive, timeless design means that »Sunny Beam« can be placed anywhere in the house, either as a table-top or as a wall-mounted unit. The unit gets its power supply from an integrated solar cell, and from a rechargeable battery for days when there is less sunlight. Via a USB interface, data can be transferred to a PC, and the battery can be recharged.

Jury
Dieses kleine Steuerungsgerät für Solaranlagen wird von Solarzellen betrieben, passt also gut ins System. Mit ihm kann ich als Besitzer einer Solaranlage präzise verfolgen, wieviel Energie meine Solaranlage gerade erzeugt. Es ist das zentrale Steuer- und Regelelement, das von der Produktgrafik her verständlich und gut gestaltet ist.
—
Judges panel
This small control unit for photovoltaic systems is itself driven by solar cells, and so matches the system nicely. It allows owners to check precisely how much power their photovoltaic systems are currently generating. It is the central control and governing unit, and its product graphics have been designed understandably and well.
—

ivo.tec

Pellet Primärofen-Kessel
Pellet-burning stove-boiler

Hersteller / Manufacturer
Wodtke GmbH
D-72070 Tübingen-Hirschau
—
Design / Designer
Joachim Pflug
D-72810 Gomaringen
—
Vertrieb / Distributor
Wodtke GmbH
D-72070 Tübingen-Hirschau
—

Mit »ivo.tec«, dem ersten Raumluft Pellet Primärofen für den Wohnraum, kann effizient und umweltschonend geheizt werden: Der Kessel wurde aus dem Keller geholt. Sein Design verbindet Ästhetik mit hohem Gebrauchswert. Purismus, zurückhaltende Eleganz, konsequente lineare Formensprache und die Verbindung von Glas und kühlem Stahl zeichnen den Primärofen-Kessel aus, der sich in jeden Wohnraum sehr gut einfügt. Der »Ofen« wird zum »Wärme-Möbel«, verbindet sich mit der Architektur des Hauses und wird zu einem Mittelpunkt der Ruhe. Die funktionale Gestaltung des Bedienboards und selbsterklärende Funktionen ermöglichen eine ergonomische Handhabung.

Heizkomfort, Bedienungsfreundlichkeit und Sicherheit sind bei »ivo.tec« zukunftsweisend. Der Betrieb ist sehr leise, der Aufwand an Reinigung und Wartung extrem gering. Das Gerät zeichnet sich durch Energieeffizienz und Schadstoffarmut aus.

»ivo.tec«, the first pellet-burning stove for living space, heats efficiently and in an eco-friendly way. It frees the boiler from the confines of the basement. Its design combines aesthetic appearance with high utility value. Puristic design, unassuming elegance, a consistent, linear formal expression, and the combination of glass and cool steel are the characteristics of this combined stove-boiler, which is not out of place in any room in the house. The »stove« now becomes a piece of »heating furniture«, harmonizes with the architecture of the house, and becomes a focal point of peace and quiet. The functional design of the operating panel and self-explanatory functions make it ergonomic to use.

»ivo.tec« shows the way forward when it comes to convenient heating, user friendliness and safety. It is very quiet in operation, and requires very little cleaning or maintenance. The appliance stands out by virtue of its energy efficiency and extremely low level of pollutant emission.

Jury
Für diese umweltfreundliche Technologie der Holzpellet-Heizungen ist es eine gut gestaltete Lösung. Die Glaselemente haben eine Sicherheitsfunktion, man verbrennt sich nicht die Finger an dem Ofen.
—
Judges panel
This is a well designed solution for environmentally friendly wood-pellet heating systems. The glass elements perform a safety function—you can never burn your fingers on the stove.
—

calimax

Bellina

Pelletofen
Pellet stove

Hersteller / Manufacturer
Calimax GmbH
A-6830 Rankweil
–
Design / Designer
Design Tech
Jürgen R. Schmid
D-72119 Ammerbuch
–

Das Heizen mit Pellets, die aus naturbelassenen Holzresten hergestellt werden, ist nicht nur ökonomisch sondern auch ökologisch sinnvoll, denn es erhöht nicht den CO_2-Gehalt in der Atmosphäre.

»Bellina« ist durch seine kompakte Bauweise verbunden mit einer hohen Heizleistung für Apartments und kleinere Wohnungen konzipiert. Er besticht durch modernes Design und frische Farben. Die abnehmbaren Seitenteile, die durch eine spezielle Beschichtung extrem widerstandsfähig sind, gibt es in acht Farben. Damit kann sich jeder seinen Ofen individuell gestalten. Mit seinen ausgewogenen, klaren Formen und seiner kompakten Erscheinung passt er sich jedem Wohnstil perfekt an.

Die innovative Steuerung ist dank ihrer einfachen Menüführung für sämtliche Funktionen äußerst übersichtlich und sehr bedienfreundlich. Aus hochwertigem Guss gefertigt ist »Bellina« sehr robust und garantiert eine lange Lebensdauer. Die Reinigung wird durch eingebaute Bürsten und Staubsaugeranschluss erleichtert.

Heating with pellets made of untreated waste timber makes not only economic, but also ecological sense, since it does not increase the level of atmospheric CO_2.

With its compact design and high heat output, »Bellina« is intended for use in flats and small houses. Its modern design and fresh colour scheme immediately catch the eye. The removable side panels, which are extremely hard-wearing due to a special coating technique, are available in eight colours. This means everyone can design their own individual stove. Its harmonious, clear forms and compact appearance make it the perfect complement to any living milieu.

The simple menu guide through all the functions of the innovative control system makes it very easy to use. Made of high-quality cast metal, »Bellina« is extremely robust, and guaranteed to have a long service life. Built-in brushes and an integrated connection point for a vacuum cleaner make cleaning easier.

Jury
Überzeugendes Konzept, klares Design, erinnert an die alten gusseisernen Öfen, einfach in der Handhabung, etwas spielerisch durch die austauschbaren Teile, die man sich farblich passend zusammenstellen kann. Kleiner Ofen, der relativ gut nachgerüstet werden kann. Passt in jedes Wohnzimmer.
–
Judges panel
Convincing concept, clear design, reminiscent of the old cast-iron stoves, simple to use, somewhat playful with its replaceable panels that can be put together to suit one's colour scheme. Small stove that can be installed relatively easily after a house has been built. Looks good in any living room.
–

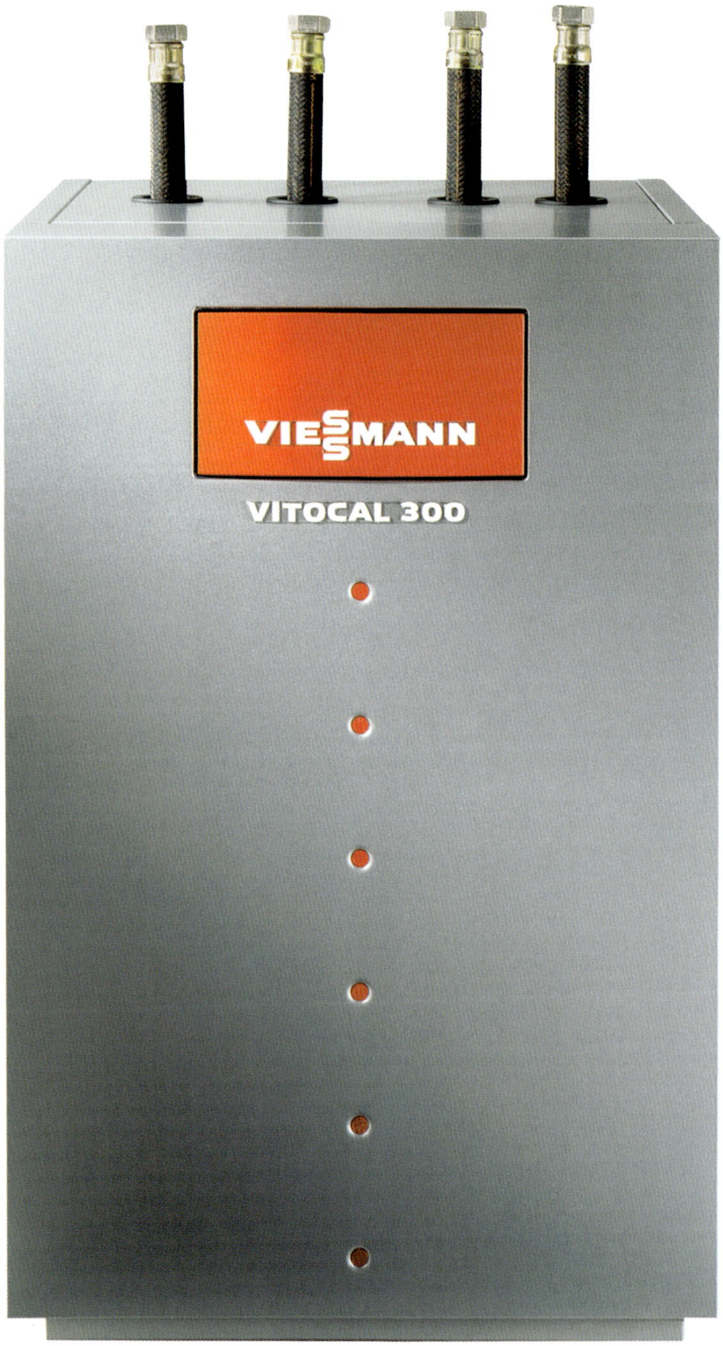
VIESSMANN
VITOCAL 300

Vitocal 300

Wärmepumpe
Heat pump

Hersteller / Manufacturer
Viessmann Werke GmbH & Co. KG
D-35107 Allendorf/Eder
—
Design / Designer
Phoenix Design
D-70376 Stuttgart
—

Die leistungsstarke und leise Wärmepumpe ist ganzjährig als vollwertiges Heizsystem nutzbar. Sie holt sich Erdwärme entweder über eine Erdsonde oder über einen Erdkollektor, der in circa 1,5 Metern Tiefe verlegt wird. Wo es möglich ist, kann aber auch mit einer Umrüstung Wärme aus dem Grundwasser genutzt werden. Mit ihrem völlig klaren Korpus und der massiven Front fügt sich die Wärmepumpe selbstbewußt in moderne Architektur.

Die menügeführte, digitale Wärmepumpenregelung überzeugt durch die anwenderfreundliche Bedienoberfläche mit großem Display und Klartextanzeige sowie integriertem Diagnosesystem. Die Regelung übernimmt das Management der kompletten Heizungsanlage und – bei Bedarf – die Regelung einer Solaranlage. Im Normalbetrieb ist sie hinter einer Klappe geschützt, im Bedienzustand herausgekippt und ergonomisch optimal von oben einzusehen und einzustellen. Mit Sonderzubehör können an heißen Tagen auch die niedrigen Temperaturen aus dem Erdreich oder Grundwasser zur Kühlung des Gebäudes genutzt werden.

This powerful yet quiet heat pump can be used all year round as a heating system in its own right. It uses heat from the ground, either via a ground probe or a ground collector at a depth of roughly 1.5 meters. Wherever possible, it can also be modified to use heat from groundwater. With its extremely plain contours and solid front panel, this heat pump can hold its own in any modern architectural setting.

The advantages of the menu-guided, digital heat-pump control are clear, with its user-friendly interface with a large screen and plain-text display, and an integral diagnostic system. The control unit takes care of the management of the entire heating system and, if necessary, of a solar heating system as well. During normal heating operation, the control unit is protected behind a flap. To operate the control unit, it is folded out, and is easy to see and adjust from above. Special accessories allow the low temperatures from the ground or the groundwater to be used to cool the building on hot days.

Jury
Sehr gut: Hier wurde der Verlockung widerstanden, vordergründiges neues Design zu erfinden. Das Produkt wurde so gestaltet, dass es sich perfekt in die gesamte Viessmann-Linie einfügt. Exzellent gestaltet. Der Zugang zum Bedienpaneel ist gut gelöst. Das Thema Erdwärme ist sicherlich im Rahmen des »Focus Energy« relevant.
—
Judges panel
Very good: they have resisted the temptation to invent superficial new design. The product has been designed so that it harmonizes perfectly with the entire Viessmann range. Excellently designed. The problem of access to the operating panel has been solved well. Ground heat is certainly a relevant topic for »Focus Energy«.
—

Jury

Eine hocheffiziente Lösung und ein solides Design. Sieht aus wie eine Pumpe aussehen soll, also insgesamt eine gute Lösung. Die Kontrollfelder sind an der richtigen Stelle, die Produktgrafik ist verständlich. Ein gutes Produkt.

—

Judges panel

A highly efficient solution and no-nonsense design. Looks like a pump should. On the whole, then, a good solution. The control panels are in the right place, the product graphics are understandable. A good product.

—

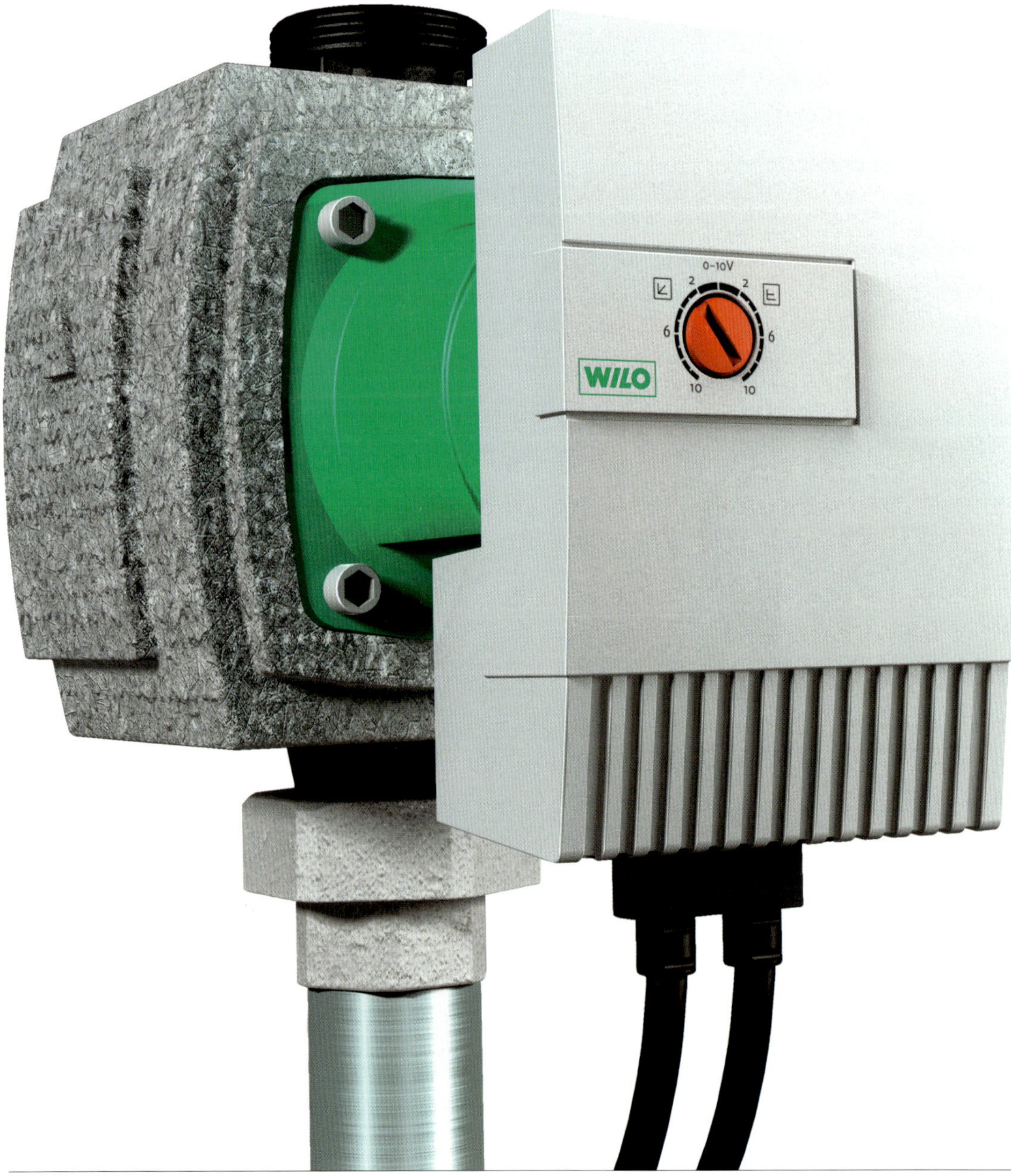

Wilo Stratos PARA

Heizungsumwälzpumpe
Heating circulation pump

Hersteller / Manufacturer
Wilo AG
D-44263 Dortmund
–
Design / Designer
Mehnert Corporate Design GmbH
D-10179 Berlin
–
Vertrieb / Distributor
Wilo AG
D-44263 Dortmund
–

Die Hocheffizienzpumpe »Wilo Stratos PARA« wird vorzugsweise in OEM-Anwendungen zur Förderung von Wasser in Heizungs- und Kühlsystemen eingesetzt. Im Vergleich zu Standardpumpen reduziert diese Pumpe den Stromverbrauch und damit die Betriebskosten um bis zu 80 Prozent. Maßgeblich hierfür ist der Einsatz einer neuen hocheffizienten EC-Motortechnologie.

Das innovative Produktkonzept setzt sich in der völlig neuartigen Anordnung der einzelnen Baugruppen fort. Damit konnten alle relevanten Einbau- und Bedienelemente dem Benutzer zugewandt und somit leicht untergebracht werden.

Die Einstellung des jeweiligen Betriebspunktes der Pumpe erfolgt über externe Signale oder anhand des roten Einstellknopfes. Die elektrische Installation über einen Stecker am Modul ermöglicht auch in schwierigen Einbausituationen einen einfachen Anschluss.

The »Wilo Stratos PARA« high-efficiency pump is mainly used in OEM applications for delivering water to heating and cooling systems. Compared with standard pumps, this pump reduces electricity consumption, and thus operating costs, by as much as 80 per cent. This is above all due to the use of a new, highly efficient, EC motor technology.

This innovative product concept is also in evidence in the completely new way the individual assemblies have been arranged. This allowed all the relevant built-in and operating elements to be positioned so that they face the user, and are thus easily accessible.

The pump's operating point is set via external signals or by means of the red selector knob. Even when there is little space for installation, a connector on the module allows easy connection to the power supply.

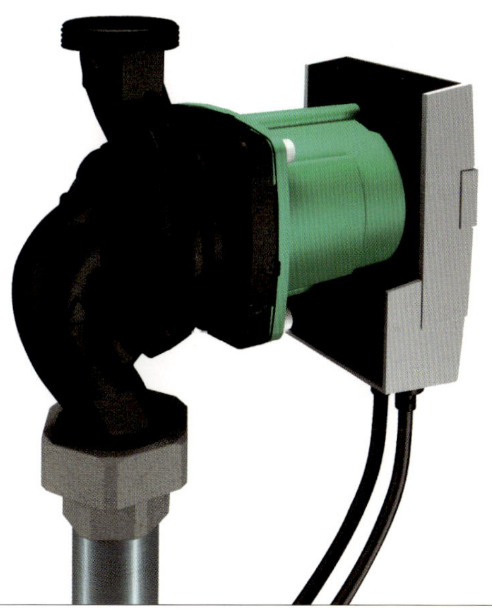

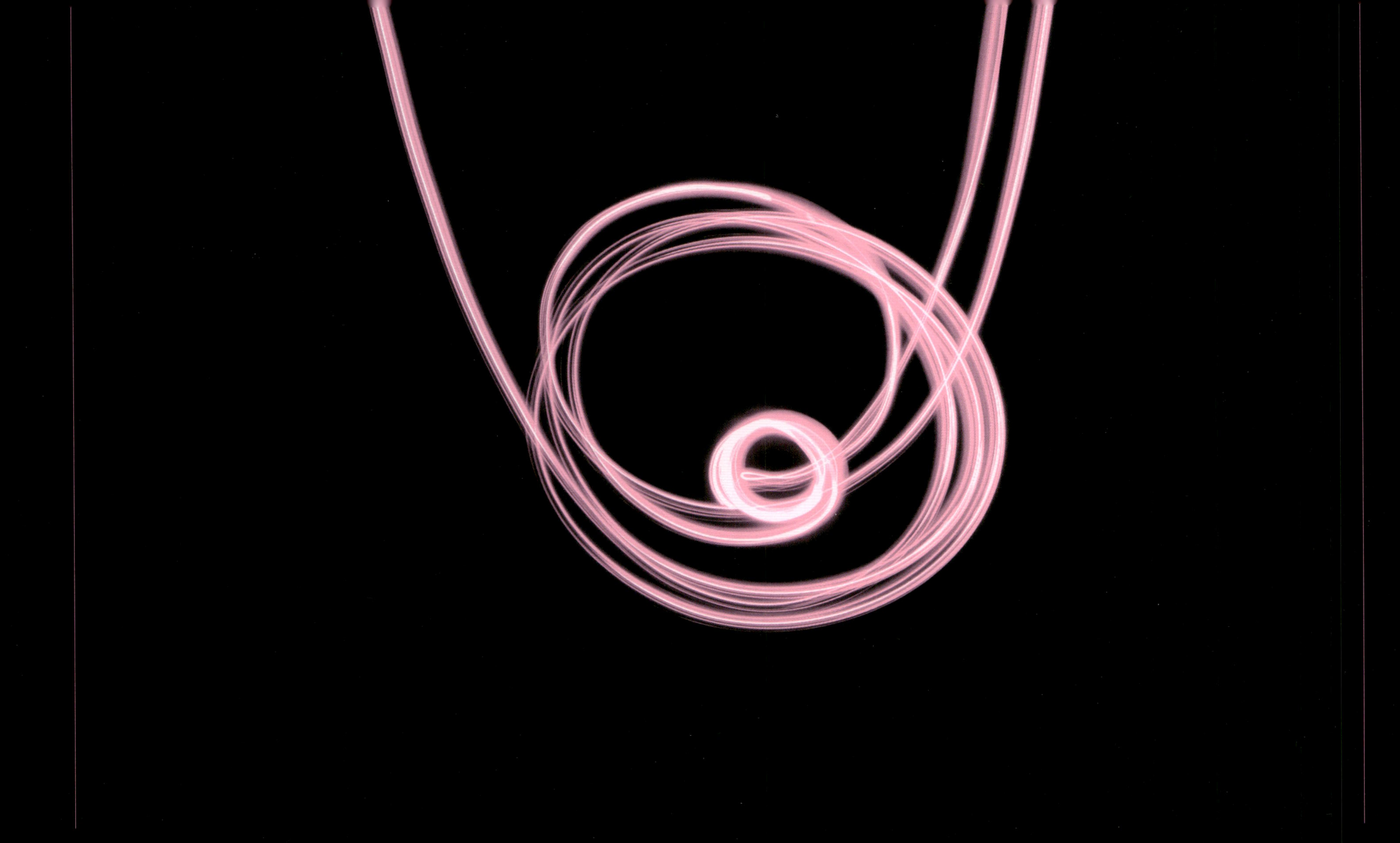

DDC 50
Bohrstaubfänger / Drilling dust collector

KÄRCHER
DDC 50

DDC 50

Bohrstaubfänger
Drilling dust collector

Hersteller / Manufacturer
Alfred Kärcher GmbH & Co. KG
D-71364 Winnenden
—
Design / Designer
Werksdesign / In-house design
Michael Meyer
—
Vertrieb / Distributor
Alfred Kärcher GmbH & Co. KG
D-71364 Winnenden
—

Bohren und zwar staubfrei – das ist mit dem Bohrstaubfänger möglich. Dank ihm muss eine zweite Person nicht mehr mit einem Staubsauger den feinen Bohrstaub, der sich oft nur schwer von der Wand wieder entfernen lässt, saugen. Mit der patentierten Vakuum-Haltefunktion saugt sich der »DDC 50« selbstständig an der Wand fest und fängt den Staub bereits beim Bohren auf – nachreinigen ist nicht mehr nötig, und man hat beide Hände für die Arbeit frei. Er kann mit jeder Bohrmaschine verwendet werden.

Der Bohrstaubfänger besteht aus einer Öffnung für den Bohrer, Behälter und Pumpe. Den Füllstand des Staubbehälters kann der Benutzer dank der transparenten Kappe leicht ablesen. Die Gestaltung fällt durch zwei kreisrunde Bereiche – Öffnung für den Bohrer und Pumpe – auf, die durch ein Trapez verbunden sind. Die zylindrische Form liegt gut in der Hand.

Das Akku-Gerät sieht nicht nur robust und kraftvoll aus, es zeichnet sich auch durch geringen Stromverbrauch aus.

The drilling dust collector makes dust-free drilling possible. Thanks to this collector, there is no need for an assistant to remove the fine drilling dust from the wall with a vacuum cleaner—often a difficult task. With its patented vacuum suction function the »DDC 50« adheres to the wall automatically, intercepting the dust during drilling. There is no need to clean up afterwards and the user has both hands free. It can be used with any power drill.

The drilling dust collector consists of an aperture for the drill bit, a dust reservoir and a pump. Thanks to the transparent cover, the user can easily see when the dust reservoir needs to be emptied. The design is striking: two circular areas—apertures for the drill bit and the pump—joined by a trapezium. The cylindrical shape fits the hand well.

Powered by a rechargeable battery, the drilling dust collector not only looks robust and powerful, but also uses little electricity.

Jury
Wunderbare Idee für den bequemen Heimwerker oder die bequeme Heimwerkerin. Wir alle kennen das Szenario: man will ein Loch bohren und braucht dafür zwei Leute – einen, der bohrt, und einen, der den Staubsauger hält. Und das geht jetzt alleine und von selbst.
—
Judges panel
A great idea for DIY enthusiasts who don't want any hassle. We're all familiar with the scenario: to drill a hole you need two people—one to do the drilling and one to hold the vacuum cleaner. And now you can do it all on your own.
—

Jury

Nicht nur ein Akkuschrauber sondern auch ein Akkubohrer. Überzeugend die sehr einfache Handhabung mit den klar erkennbaren Verstellmöglichkeiten und guter Produktgrafik. Das Produkt ist schwer, weil es sehr stark und robust ist, liegt aber trotzdem gut in der Hand.

—

Judges panel

Not only a cordless screwdriver, but also a cordless drill. The very easy handling, plainly recognizable setting options and good product graphics are impressive. The product is heavy because it is very rugged and powerful, but it is nonetheless comfortable to hold.

—

BBS Raptor

Akkuschrauber
Cordless drill-driver

Hersteller / Manufacturer
AEG Elektrowerkzeuge
D-71364 Winnenden
—
Design / Designer
Kienledesign
D-73575 Leinzell
—

Die BBS Raptor-Serie umfasst vier besonders kraftvolle Akkubohrschrauber für professionelle Anwender in Handwerk und Industrie. Die neuen Geräte vereinen starke Leistung mit einer sehr kompakten und ergonomischen Bauform. In Verbindung mit dem patentierten Wendeakkusystem und einem leicht abnehmbaren Bohrfutter sind Anwender mit diesen Geräten besonders bei Überkopfarbeiten und bei Arbeiten an schwer zugänglichen Stellen im Vorteil.

Dank des neu entwickelten Super-Fixtec-Systems kann das Bohrfutter in Sekundenschnelle entfernt werden. Das verkürzt die Maschine um 25 Prozent. Dadurch verbessert sich die Handlichkeit bei beengten Platzverhältnissen. Der Ablösemechanismus ist äußerst robust und sehr einfach zu bedienen. Noch mehr Bewegungsfreiheit und eine stets optimale Balance des Gerätes wird dadurch erreicht, dass der Akku um 180 Grad gedreht eingeschoben werden kann.

Robuste Metallgetriebekästen mit besonders kurzer Bauform, selbstreinigende Akkukontakte und der ergonomische T-Handgriff mit Softgrip zur Verminderung von Vibrationen unterstreichen den professionellen Anspruch.

The BBS Raptor series includes four especially powerful cordless drill-drivers for professional users in the craft trades and industry. These new tools combine powerful performance with a very compact and ergonomic design. In combination with the patented reversible battery system and an easily detachable chuck, users of these tools benefit in particular when working overhead and in hard-to-reach places.

Thanks to the new Super-Fixtec system, the chuck can be removed within seconds, making the tool 25 per cent shorter, which improves handling in confined spaces. The release mechanism is extremely robust and very easy to operate. The fact that the battery can also be inserted after turning it through 180 degrees gives even more freedom of movement and ensures that the tool is always optimally balanced.

Robust, very short metal drive housings, self-cleaning battery contacts and the ergonomic soft-grip T-handle to reduce vibrations emphasize the professional standard.

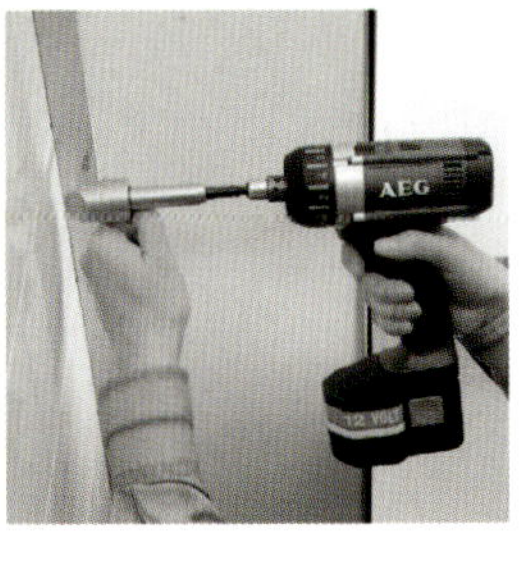

Wiha Professional
Wiha Professional Electric

Wiha Professional

Kombinationszange
Combination pliers

Hersteller / Manufacturer
Wiha Werkzeuge GmbH
D-78136 Schonach
—
Design / Designer
Ahape – more than design
Ahape GmbH
Eckhard Ruoff und Herbert Pauser
D-64283 Darmstadt
—

Die Kombinationszange »Wiha Professional« ist eine Universalzange und bietet sicheren Halt beim Greifen, Biegen sowie Schneiden weicher und harter Drähte und Kabel. Sie besitzt eine extra lange Schneide für Flach- und Rundkabel. Gezahnte Greifbacken und der breite Griffrücken der neuen, ergonomischen Mehrkomponenten-Griffhülle bieten selbst bei kraftbetontem häufigem und langem Arbeiten höchsten Komfort.

Die komfortablen, mehrkomponentigen Griffe sind aus unterschiedlich weichen, elastischen, rutschfesten und schlagdämpfenden Kunststoffen, die je nach Druck- oder Bewegungszonen optimal verteilt sind. Das sorgt für ermüdungsarmes Arbeiten und optimale Übertragung der Handkraft ohne Druckstellen.

Der stabile, ergonomisch geformte Griff-Grundkörper ist fest mit der Zange verbunden. Die ausgeformten Griffenden mit Anschlag verhindern ein Abrutschen. Das harmonische vordere Griffdesign bietet optimalen Halt der Finger bei kraftbetonter Anwendung.

Dazu kommt eine hochwertige Mehrschicht-Verchromung für perfekten dauerhaften Korrosionsschutz. Die Schneiden sind zusätzlich induktiv gehärtet. Durch Verwendung spezieller Nietqualität ist das Gelenk äußerst verschleißarm und hochbelastbar.

The »Wiha Professional« are all-purpose combination pliers for securely gripping, bending and cutting both hard and soft wires and cables. They have an extra-long blade, suitable for ribbon and round-section cables. Toothed jaws and the wide backs of the new, ergonomic, multi-component grip handle sleeves ensure outstanding comfort even when doing strenuous work for prolonged periods of time.

The comfortable, multi-component grip handle sleeves are made of unequally soft, elastic, non-slip, shock-absorbing plastics which are optimally distributed according to areas of pressure or movement. This helps ensure that using the pliers is less tiring and that force is optimally transferred without bruising.

The sturdy, ergonomically shaped basic handle elements are permanently connected to the pliers. The shaped ends of the handles, with stops, prevent the fingers from slipping. The harmonious design of the front of the handles provides optimal support for the fingers when applying force.

High-quality laminar chrome-plating ensures perfect, durable corrosion protection. In addition, the blades are inductively hardened. Thanks to special-quality riveting, the joint is extremely wear-resistant and robust.

Jury
Überzeugend die Ergonomie, die Materialwahl und auch die Möglichkeit des Verschließens. Optimale Kraftübertragung.
—
Judges panel
Convincing the ergonomics, the choice of the materials and the fact that the pliers can be locked. Optimal transmission of forces.
—

Jury

Klassische Form der Schere durch das Material neu interpretiert. Sehr ergonomisch, für Textilverarbeitung mit klebriger Oberfläche. Schere ist unempfindlich gegen aggressive Reinigungsmittel, sie kann in einem Acetonbad gereinigt werden.

—

Judges panel

Classical scissors design re-interpreted with new materials. Highly ergonomic, for cutting textiles with adhesive surfaces. The scissors are also resistant to aggressive cleaning agents and can be cleaned in an acetone bath.

—

Serie D 900

Hightech-Scheren
High-tech scissors

Hersteller / Manufacturer
Bessey Tool GmbH & Co. KG
D-74301 Bietigheim-Bissingen
—
Design / Designer
Weinberg & Ruf
D-70794 Filderstadt
—
Vertrieb / Distributor
Bessey Tool GmbH & Co. KG
D-74301 Bietigheim-Bissingen
—

Die »Serie D 900« besteht aus drei völlig neu entwickelten Hightech-Scheren der Marke Erdi. Die »Fibre« für Glasfasern und Kohlefasern hat keine Verzahnung. Dadurch sind lange schiebende Schnitte möglich, wenn zum Beispiel für Windkraftanlagen Rotorblätter hergestellt werden.

Ideal für kurze und lange Schnitte bei Glas- und Kohlefasern ist die »Fibre Z« mit Mikroverzahnung. Das verhindert, dass engmaschiges und glattes Gewebe abrutscht. Die Schneiden der »Aramid« haben eine Härte von 60 bis 62 HRc und sind zusätzlich hartbeschichtet. Deshalb ist sie auch ideal für die zähen Aramidfasern, die beispielsweise im Boots-, Flugzeug- und Rennwagenbau oder für Schutzkleidung der Polizei oder der Feuerwehr verwendet werden.

Dank des Präzisionsschliffes gleiten alle Scheren sauber und leichtgängig durch die Fasern. Das Material muß jetzt nicht mehr wie bisher angehoben werden, da die Griffaugen in einem steileren Winkel zu den Klingen angeordnet sind. Die Scheren können nun glatt über die Arbeitsfläche geführt werden ohne die sonst übliche störende Schaukelbewegung. Sie ermöglichen kraftvolles und bequemes Arbeiten. Die Griffaugen sind so groß, dass man auch mit Handschuhen arbeiten kann, ohne Druckstellen an den Händen zu bekommen.

The »D 900 Series« consists of three completely new pairs of high-tech scissors, marketed under the Erdi brand name. The »Fibre« scissors for glass and carbon fibre are not serrated. This makes long sliding cuts possible, for example when making rotor blades for wind turbines.

For both short and long cuts in glass and carbon fibres, the micro-serrated »Fibre Z« scissors are ideal. The micro-serration prevents close-meshed and smooth fabrics from slipping away. The blades of the »Aramid« scissors are hardened to 60–62 HRC and are additionally hard-coated. This also makes them ideal for cutting the tough Aramid fibres used, e.g., in the construction of yachts, aircraft and racing cars, and for protective clothing for policemen and fire-fighters.

Thanks to high-precision grinding all the scissors slide through the fibres smoothly and cleanly. Also, the material no longer has to be raised, because the finger-holes are at a steeper angle relative to the blades. So the scissors can be moved smoothly across the work surface without the usual rocking motion, thus making it possible to work energetically and comfortably. The finger-holes are large enough to allow users to wear gloves without getting pressure sores on their hands.

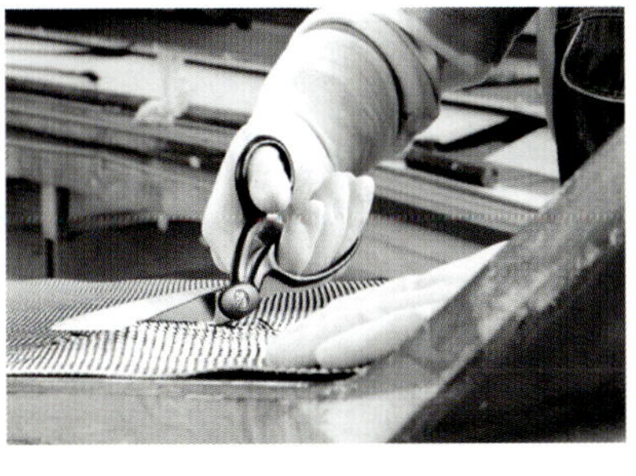

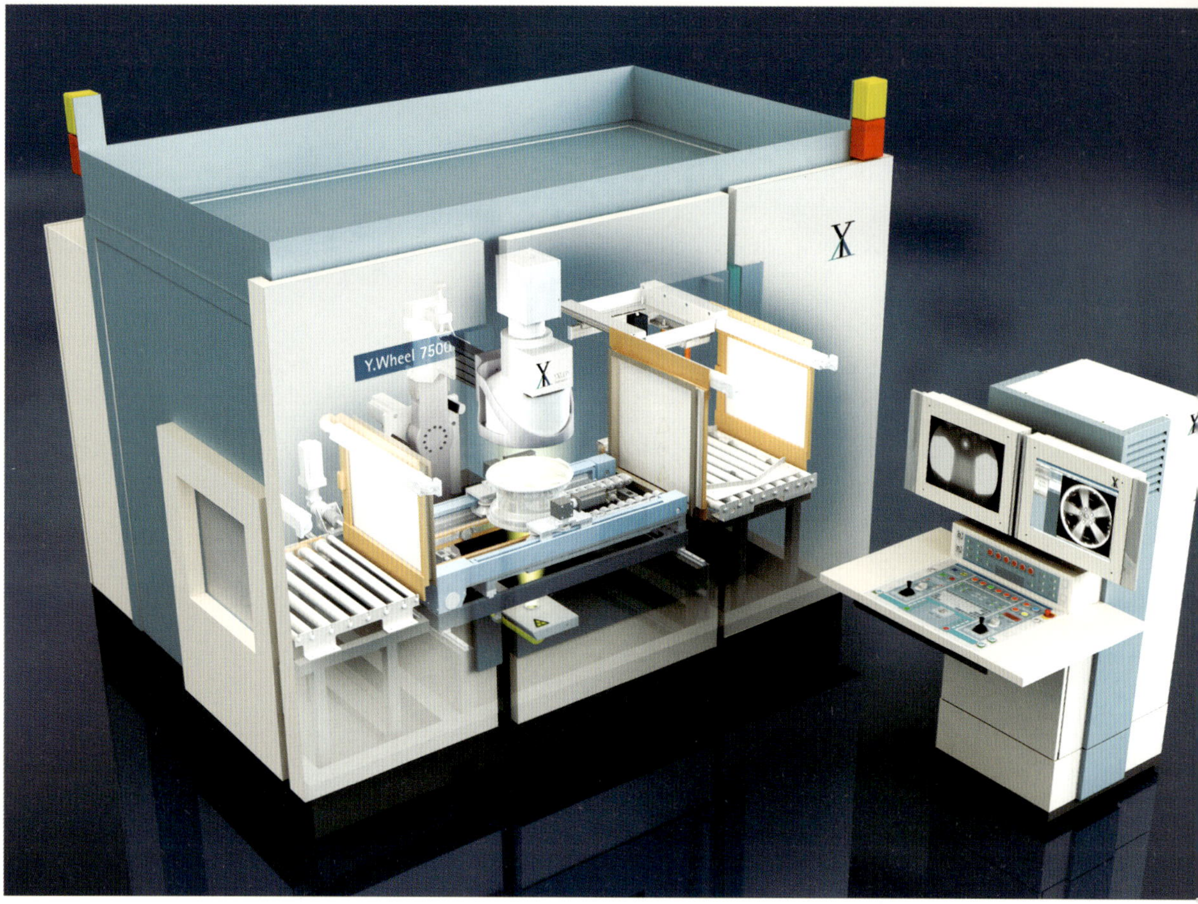
Y.Wheel 7500

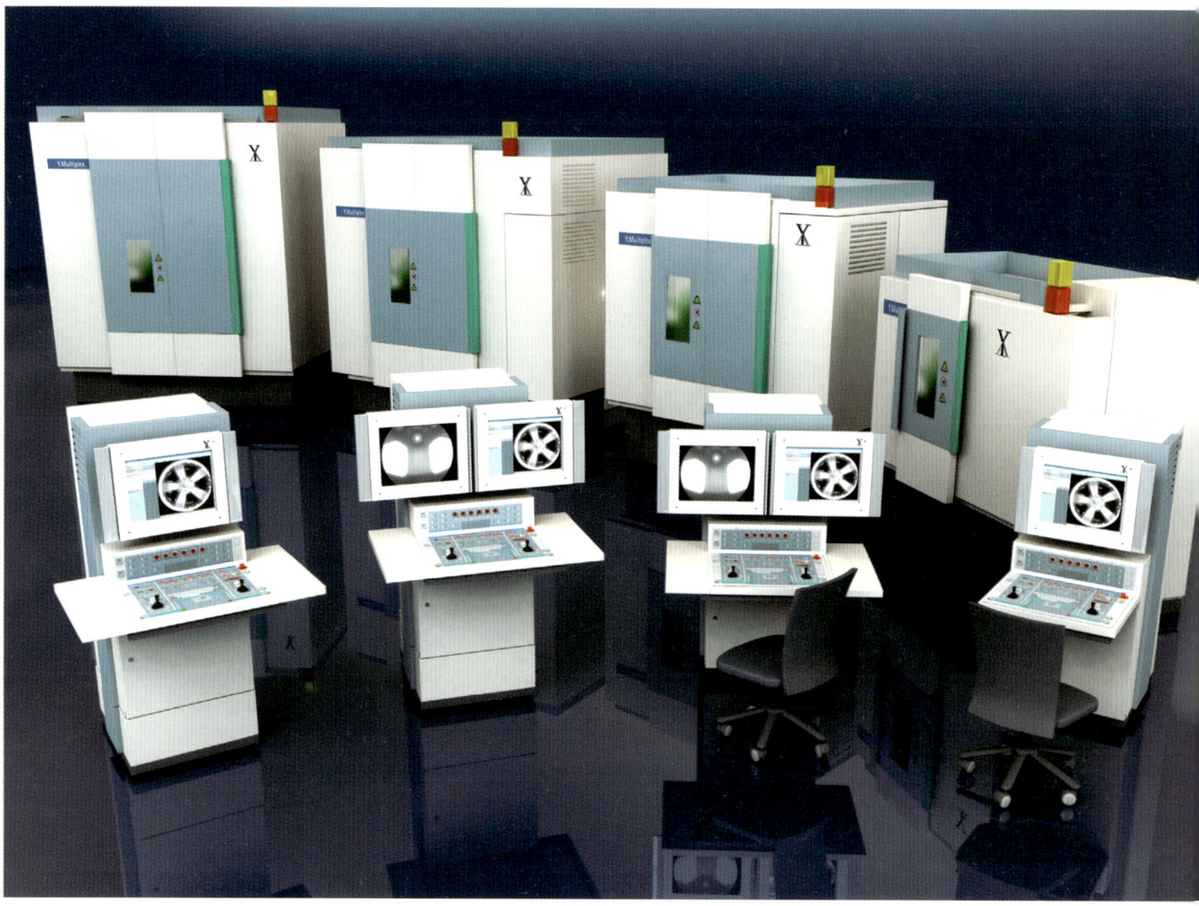

YXLON.Wheel
YXLON.Multiplex
YXLON.Carrier

Prüfsysteme
Testing systems

Hersteller / Manufacturer
YXLON International X-Ray GmbH
D-22419 Hamburg
—
Design / Designer
BGP Design
D-70563 Stuttgart
—
Vertrieb / Distributor
YXLON International X-Ray GmbH
D-22419 Hamburg
—

Die YXLON.Produkte sind veschiedene Prüfsysteme, die sich durch ein einheitliches Corporate Design auszeichnen, das weit über das reine Produktdesign hinausgeht. Alle haben eine kompakte Bauform, benötigen nur geringe Stellflächen und lassen sich einfach integrieren. Das Systemdesign besticht durch seine klare und funktionale Gestaltung.

»YXLON.Multiplex«, ein modulares Prüfsystem für die universelle Materialprüfung, lässt sich gleichermaßen gut für die Stichproben- und Serienprüfung einsetzen und bietet hochwertige Prüfqualität bei so unterschiedlichen Materialien wie Stahl, Aluminium, Keramik, Kunststoff oder Gummi.

Die neue »YXLON.Carrier« für die palettenbasierte, automatische Gussteilprüfung ermöglicht durch hervorragende Bildqualität auch bei komplexen Strukturen eine eindeutige Analyse – und dies bei kurzen Prüfzeiten.

Die automatischen Räderprüfsysteme »YXLON.Wheel« lassen sich sehr flexibel, auch im automatischen Betrieb, sicher und schnell an Prüfspezifikationen unterschiedlicher Kunden anpassen.

The YXLON.products are various test systems featuring a uniform corporate design that goes far beyond mere product design. All are compact, have a small footprint and are easy to integrate. Clear functionality is an impressive characteristic of the overall system design.

»YXLON.Multiplex«, a modular system for universal materials testing, is equally appropriate for random and series testing, with high-quality testing of materials as diverse as steel, aluminium, ceramics, plastics and rubber.

Thanks to excellent image quality, the new »YXLON. Carrier« for pallet-based automated inspection of castings allows unequivocal analisis even with short inspection times.

The »YXLON.Wheel« automatic wheel testing systems can be adapted quickly and reliably, and even automatically, to the test specifications of different customers.

Jury
Gutes Beispiel für rationales Corporate Design in einem schwierigen Feld. Sehr gutes Industriedesign, bestechende Klarheit. Überzeugend der Corporate Design- und Systemgedanke, der sicher in hohem Maße auch Rationalisierung bedeutet.
—
Judges panel
Good example of rational corporate design in a difficult area. Very good industrial design, excellent clarity. The corporate design and the system philosophy, which undoubtedly also spell »rationalization«, are convincing.
—

KÄRCHER
CB line

KÄRCHER
CB line

CB Line

Autowaschanlage
Gantry car wash

Hersteller / Manufacturer
Alfred Kärcher GmbH & Co. KG
D-71364 Winnenden
—
Design / Designer
Werksdesign / In-house design
Denis Dammköhler
—
Vertrieb / Distributor
Alfred Kärcher GmbH & Co. KG
D-71364 Winnenden
—

Die »CB Line«, ein innovatives Anlagenkonzept vom Einsteigermodell bis zum professionellen Waschcenter, ist das Ergebnis einer umfassenden Neuentwicklung, bei der mit durchdachter Technik und unter Verwendung hochwertiger Komponenten ein einfaches, wartungsfreies System in einem attraktiven Erscheinungsbild geschaffen wurde.

Sie zeichnet sich durch einen sparsamen Umgang mit Ressourcen aus. So reduziert Recycling-Wasser den Frischwasserverbrauch um bis zu 85 Prozent. Dazu wird das Waschwasser über eine physikalische Filtration so aufbereitet, dass es dem Waschprozess wieder vollständig zugeführt werden kann. Der Frischwasserverbrauch reduziert sich auf das Klarspülen sowie die Zugabe von Pflegeprodukten.

Und dank der Beleuchtung durch LEDs, die eine 50-mal höhere Lebensdauer haben, wird deutlich weniger Strom verbraucht. Außerdem sorgt der optimierte Dachtrockner mit weniger Leistung, rund 75 Prozent, für bessere Trocknungsergebnisse als herkömmliche Modelle. Die Anlagen sind einfach zu bedienen.

The »CB Line«, a novel car wash concept, available in several versions from an entry-level model to a professional wash centre, is a completely new development. With well thought-out technology and using high-quality components, a simple, maintenance-free system has been created, with an attractive appearance.

One of its main attributes is its low consumption of resources. The use of recycled water, for instance, reduces its fresh water consumption by up to 85 per cent. To this end the wash water is treated by physical filtration in such a way that it can be returned to the wash process without loss. Fresh water consumption is thus reduced to rinsing and the addition of care products.

And thanks to lighting by LEDs, which have a 50-fold higher lifetime, electricity consumption is significantly lower. In addition, the optimized roof dryer delivers better results than conventional models with only 75 per cent of their power requirement. The washes are easy to operate.

Jury
Überzeugend die ökologischen Aspekte. Wasser wird durch das Recycling des Waschwassers eingespart. Unabhängig von der guten formalen Gestaltung ist das ein überzeugendes Argument bei einem Designpreis mit dem Schwerpunkt »Energy«. Sanftes Waschverfahren durch weiche Bürsten.
—
Judges panel
The ecological aspects are impressive. Water is saved by recycling the wash water. Independently of the good design, that is a convincing argument in a design competition focusing on energy. Gentle wash process thanks to soft brushes.
—

VOITH
VOITH

Voith Papiermaschine

Maschine zur Herstellung grafischer Papiere

Graphic paper machine

Hersteller / Manufacturer
Voith Paper GmbH & Co. KG
D-89522 Heidenheim
—
Design / Designer
Designpraxis Diener
D-89079 Ulm
—
Vertrieb / Distributor
Voith Paper GmbH & Co. KG
D-89522 Heidenheim
—

Das Plattformkonzept bietet aus ökologischer Sicht erhebliche Vorteile. Die Papiermaschine lässt sich dank des modularen, geschraubten Aufbaus komplett rückbauen und weiter verwenden. So kann die Maschine an andere Standorte verlagert werden, was besonders beim Weiterverkauf oder Umbau einer Anlage wichtig ist. Das Ressourcen sparende Baukastensystem erlaubt je nach Kundenwunsch und produzierter Papierart einzelne Systemkomponenten modular zusammenzustellen.

Die Maschinenbereiche sind verkapselt, dadurch kann ein interner weitgehend geschlossener Energie- und Wasserkreislauf aufrecht erhalten werden. So wird zum Beispiel die entstehende Wärmeenergie zur Trocknung des Papiers verwendet.

Die Papiermaschine zeichnet sich durch bis ins Detail gehende Gestaltung, einheitliche Benutzerführung und dynamisches Aussehen trotz beträchtlicher Maschinengröße aus. Das innovative Designkonzept kommuniziert die herausragende Produktqualität, die Verlässlichkeit der Maschine und deren hohe Arbeitsgeschwindigkeit.

From an ecological viewpoint, the platform concept offers substantial advantages. Thanks to the modular, bolted construction, the paper machine can be completely disassembled and re-used. In this way the machine can be transferred to other locations, which is important in particular if it is resold or converted. The resource-saving unit construction system allows system components to be individually combined according to the customer's requirements and the type of paper produced.

The machine sections are sealed, so that almost fully closed internal energy and water circuits can be maintained. For example, the thermal energy generated is used for drying the paper.

Features that set this paper machine apart are consistency of the design right down to small details, the uniform user prompting and the machine's dynamic appearance, despite its considerable size. The innovative design concept communicates the outstanding product quality, the reliability of the machine and its high productivity.

Jury

Hervorragendes Beispiel für ausgezeichnetes Investitionsgüterdesign. Gegenüber dem Vorgängermodell wurde die Teilevielfalt reduziert, die verwendeten Materialien sind energiesparender, leichter, bessere Lagerungen von Walzen, das Trocknungsverfahren für die nassen Papierbahnen wurde optimiert. Mehr Energie kann eingespart werden, und in puncto Arbeitssicherheit ist die Maschine übersichtlicher und durch bessere Verkleidung sicherer geworden.
—

Judges panel

Outstanding example of excellent capital equipment design. As compared to the previous model, the variety of parts has been reduced, the materials used are lighter and more energy-saving, the rollers have improved bearings and the drying process for the wet paper webs has been optimized. Thus more energy can be saved, and as regards working safety the machine is now more clearly arranged and has been made safer by improved cladding.
—

ICPS

Retortenmeiler
Retort charcoal kiln

Hersteller / Manufacturer
Adam + Partner
D-82467 Garmisch
—
Design / Designer
Adam + Partner
D-82467 Garmisch
—
Vertrieb / Distributor
Adam + Partner
D-82467 Garmisch
—

In den sogenannten Südländern wird Haushaltsholzkohle meist mit einer umweltschädlichen und ineffektiven Erdhügel-Methode aus dem Mittelalter hergestellt. Dafür wurde nun ein innovativer Retortenmeiler »ICPS« – auch »adam-retort« genannt – mit einem Zwei-Phasensystem entwickelt, der mit lokalen Baumaterialien und geringen Kosten gebaut werden kann und damit große Chancen hat, von der Bevölkerung angenommen zu werden.

Beim traditionellen Verfahren belasten Rauchschwaden vier bis sechs Tage die Umwelt. Beim »ICPS« dagegen werden sie in der Feuerkammer verbrannt, und die ganze Verkohlung dauert gerade mal 10 Stunden. Die Effektivität liegt bei 35 bis 40 Prozent, beim alten Verfahren nur bei 12 bis 22 Prozent.

Zunächst wird das Holz in der Retortenkammer durch heiße Rauchgase, die mit Abfallholz in einer externen Feuerkammer entstehen, getrocknet und die Verkohlung eingeleitet. Sobald das Wasser im Holz verdampft wurde und die ersten brennbaren Holzgase entstehen, wird der Rauch umgeleitet und in der heißen Feuerkammer umweltfreundlich abgefackelt. Das ermöglicht, dass gutes Holz nicht mehr wie bisher verbrannt werden muss.

In developing countries, household charcoal is usually produced by an ineffective, medieval mound method, which causes environmental pollution. Now an innovative retort kiln, the »ICPS« or »adam-retort« has been developed, with a two-phase system. It can be built with locally available materials and at low cost, and thus stands a very good chance of being accepted.

With the traditional method, clouds of smoke pollute the environment for four to six days. By contrast, in the »ICPS« they are burned in the combustion chamber and the entire carbonization process takes just 10 hours. Its efficiency is 35 to 40 per cent, while that of the old method is only 12 to 22 per cent.

First, the wood is dried in the retort chamber by hot flue gases generated with waste wood in an external combustion chamber. As soon as the water in the wood has evaporated and the first combustible wood gases are produced the smoke is diverted and burned off in an eco-friendly manner in the hot combustion chamber. This means that it is no longer necessary to burn good wood.

Jury
Wunderbarer Ansatz, für sogenannte Entwicklungsländer zu gestalten. Und zwar ist es hier kein fertiges Objekt, das dorthin transportiert wird, sondern es ist eine Hilfe zur Selbsthilfe. Also eine Anlage, die Menschen befähigt, eine eigene ökonomische Aktivität zu entwickeln. Hier ist zudem die Energieeffizienz der Produktion gravierend erhöht worden im Vergleich zur traditionellen Methode. Insgesamt eine runde und sehr nachhaltige Lösung.
—

Judges panel
Fine attempt to design for so-called developing countries. And in this case it is not a finished article that is transported there, but help to self-help. That is, a system that puts people in a position to develop an economic activity of their own. Moreover, the energy efficiency of production has been vastly increased as compared to the traditional method. All in all a complete and very durable solution.
—

Lernspielzeug Energie
Technisches Lehrmittel / Technology teaching aid

Kommunikation, Information / Gold
Communication, information / Gold

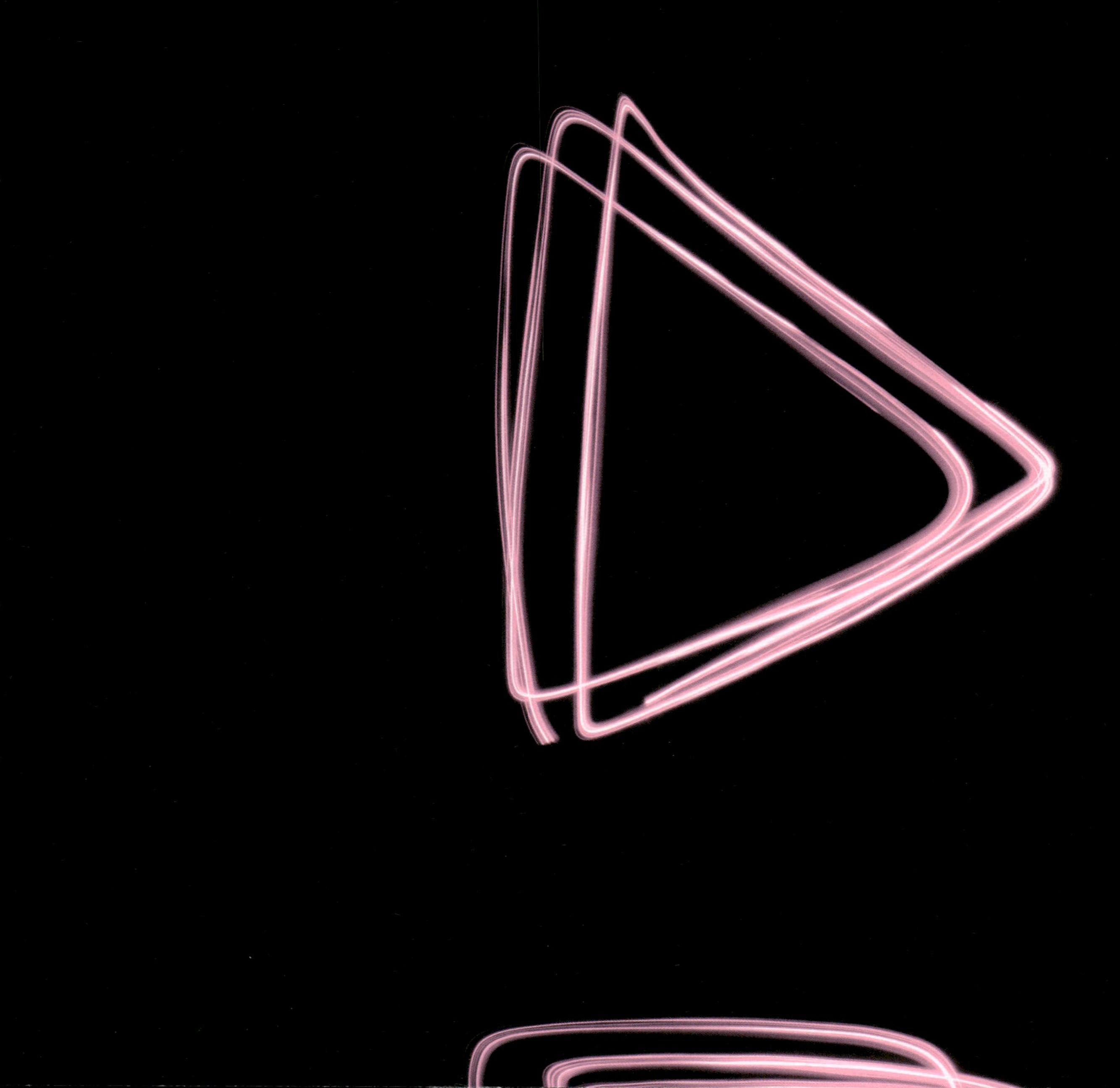

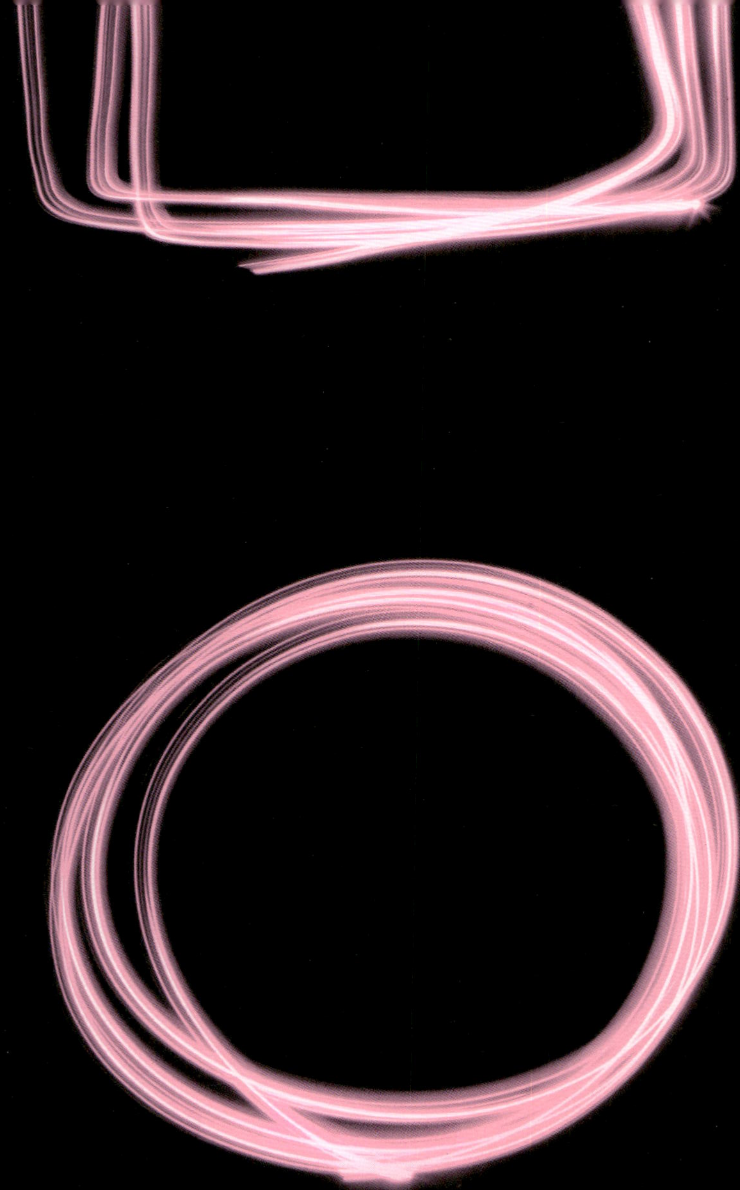

Jury

Lernspiel mit sehr viel Experimentierelementen für Kinder und Jugendliche, die den Zusammenhang von Elektrik, Solarenergie und manuell betriebener Energie selbst ausprobieren können. Modulares System mit einfachen Elementen, leicht zu verstehen, besitzt einen hohen Aufforderungscharakter. Gut gelöst, sehr ansprechend gestaltet.

—

Judges panel

Educational toys with a large number of components for experiments, to enable children and adolescents to explore the relationships between electricity, solar energy and manually generated energy for themselves. Modular system with simple components, easy to understand, cries out to be used! Well conceived, very attractively designed.

—

Lernspielzeug Energie

Technisches Lehrmittel
Technology teaching aid

Hersteller / Manufacturer
Jürgen Breitkopf Products
D-48161 Münster
–
Design / Designer
Jürgen Breitkopf Products
D-48161 Münster
–
Vertrieb / Distributor
Jürgen Breitkopf Products
D-48161 Münster
–

Das Lernspielzeug, das für Schulen, freie Bildungseinrichtungen und den Freizeitbereich konzipiert ist, soll Kinder und Jugendliche für das Thema Energie, speziell der elektrischen Energie, sensibilisieren.

Kinder erfahren dabei spielerisch, wie elektrische Energie durch erneuerbare Energien, wie Photovoltaik oder Windräder, gewonnen wird. Daneben gibt es einen Handkurbelgenerator, mit dem per Muskelkraft die gleiche Leistung erzeugt wird, wie mit Windrad oder Photovoltaik. Als Verbraucher dienen Produkte, die Kinder aus ihrem Alltag kennen: elektrische Kleingeräte wie Radio, Walkman oder ein Legomotor.

Auf der Verbraucherseite wird durch Vergleiche zwischen herkömmlichen und neuen Technologien effiziente Energienutzung anschaulich, zum Beispiel im Vergleich zwischen Glühlampen und weißen LEDs. Antriebe (Motoren), die für sich betrieben werden können oder mit Adapter in das Legosystem passen, zeigen, wie durch entsprechende Technologie mit Energie umgegangen werden kann. Elektrische Kleinverbraucher, wie ein Taschenradio, können problemlos betrieben werden.

An einem Modellhaus mit elektrisch funktionierendem Herd, Waschmaschine und Beleuchtung kann vermittelt werden, was große und kleine Verbraucher im Haushalt sind. Der Stromverbrauch der einzelnen Geräte im Verhältnis zueinander entspricht dem der Realität. Die gelungene Produktgestaltung trägt dazu bei, das Thema durch Sinnlichkeit, Anschaulichkeit und Klarheit attraktiv darzustellen.

These educational toys, designed for schools, independent educational institutions and the leisure sector, are intended to promote awareness of energy, in particular electrical energy, among children and adolescents.

Using the toys, children learn through play how electrical energy can be generated with renewable energy, e.g. photovoltaics or wind turbines. The set includes a hand crank generator which generates the same power output as a wind turbine or solar panels. The consumers are products that children are familiar with – small electrical appliances such as a radio, a Walkman or a Lego motor.

On the consumer side, efficient use of energy is demonstrated by comparisons between conventional and new technologies, e.g. between incandescent bulbs and white LEDs. Drive units (motors) that can be operated on their own or – by means of an adapter – used with the Lego system, show how energy can be handled with appropriate technology. Small appliances such as a pocket radio can easily be powered.

By means of a model house with a working electric cooker, washing machine and lighting the difference between large and small domestic electrical consumers can be illustrated. The power consumption of the individual appliances relative to one another corresponds to reality. The good product design helps to make the topic attractive – via the senses, pleasant appearance and clarity.

Jury
Gut gestaltetes Produkt für eine sehr sinnvolle Anwendung: die Ausstattung von Schulen oder anderen Einrichtungen mit Laptops. Dank der Mobilität und Flexibilität kann man viele Räume und feste Installationen sparen. Die Geräte sind somit auch vor Vanadalismus sicher. Gut eingesetztes Design, intelligent für einen sozial sinnvollen Zweck.
—
Judges panel
Well-designed product for a very sensible application—equipping schools or other institutions with laptops. Thanks to its mobility and versatility it makes a large number of rooms and permanent wiring etc. superfluous, as well as protecting the computers from vandalism. Well applied design, intelligent and for a socially desirable purpose.
—

Notebook Mobil™

Transport und Aufbewahrung für Notebooks

Transport and storage for notebooks

Hersteller / Manufacturer
Job Factory AG
CH-4057 Basel
—
Design / Designer
Design-Werkstatt Weizenkorn
CH-4123 Allschwil
—
Vertrieb / Distributor
Job Factory AG
CH-4057 Basel
—

Beim Einsatz von IT-Infrastruktur in Schulen, Messezentren und anderen Institutionen verändert sich viel, festinstallierte Geräte in Schulungsräumen gehören der Vergangenheit an. Das »Notebook Mobil™« bietet hierfür eine innovative und flexible Lösung, um Notebooks aufzubewahren und zu transportieren. Es übernimmt sämtliche Möglichkeiten eines Informatikraums, ersetzt aufwändige Rauminstallationen und erhöht die Raum- und Geräteauslastung.

Das »Notebook Mobil™« verfügt über ein Docking-System für bis zu 20 Steckplätze, die an alle gängigen Notebooks angepasst werden können. Die Notebooks werden über ein zentrales Netzgerät sicher geladen. Dank einer optimalen Luftzirkulation im Mobil wird die entstehende Wärme ohne Lüfter abgeführt. Peripheriegeräte, wie Drucker oder Scanner, können in den Stauräumen und Schubladen platziert werden. Innovative hochwertige Werkstoffe, gute Verarbeitung und robuste Konstruktion erlauben einen täglichen und flexiblen Einsatz in Schulen und anderen Institutionen. Das moderne benutzerorientierte Design besticht durch seine Funktionalität.

A lot is changing in the way IT infrastructure is being used in schools, exhibition centres and other institutions. Permanently installed equipment in training rooms is a thing of the past. A novel and flexible solution to the problem of storing and transporting notebook PCs is the »Notebook Mobil™« trolley. It assumes all the functions of a computer room, replacing complex and costly IT installations and increasing the utilization of room and equipment capacity.

The »Notebook Mobil™« has a docking system for up to 20 slots, which can be adapted to all standard notebooks. The notebooks are reliably charged by a central mains power supply unit. Thanks to optimal air circulation in the trolley, the heat generated is dissipated without a fan. Peripherals such as printers or scanners can be put in the storage spaces and drawers. Made of new, high-quality materials, robust and well-finished, the versatile »Notebook Mobil™« is built for everyday use in schools and other institutions. Functionality is an outstanding feature of the modern, user-oriented design.

Jury

Klare und saubere Lösung für ein Produkt, bei dem man als Investitionsgut kein Design erwartet. Es ist trotzdem eine sehr aufgeräumte klare Lösung, die es fast in die Nähe eines Designobjekts bringt. Sehr sicher für die Polizei zum Beispiel. Gute Designlösung.

—

Judges panel

Clear, neat solution. With products of this kind—as capital investments—design is not expected. Nevertheless, a very tidy, well-organized and clear solution that brings it close to the status of a design product. Very secure, e.g. for the police. Good design.

—

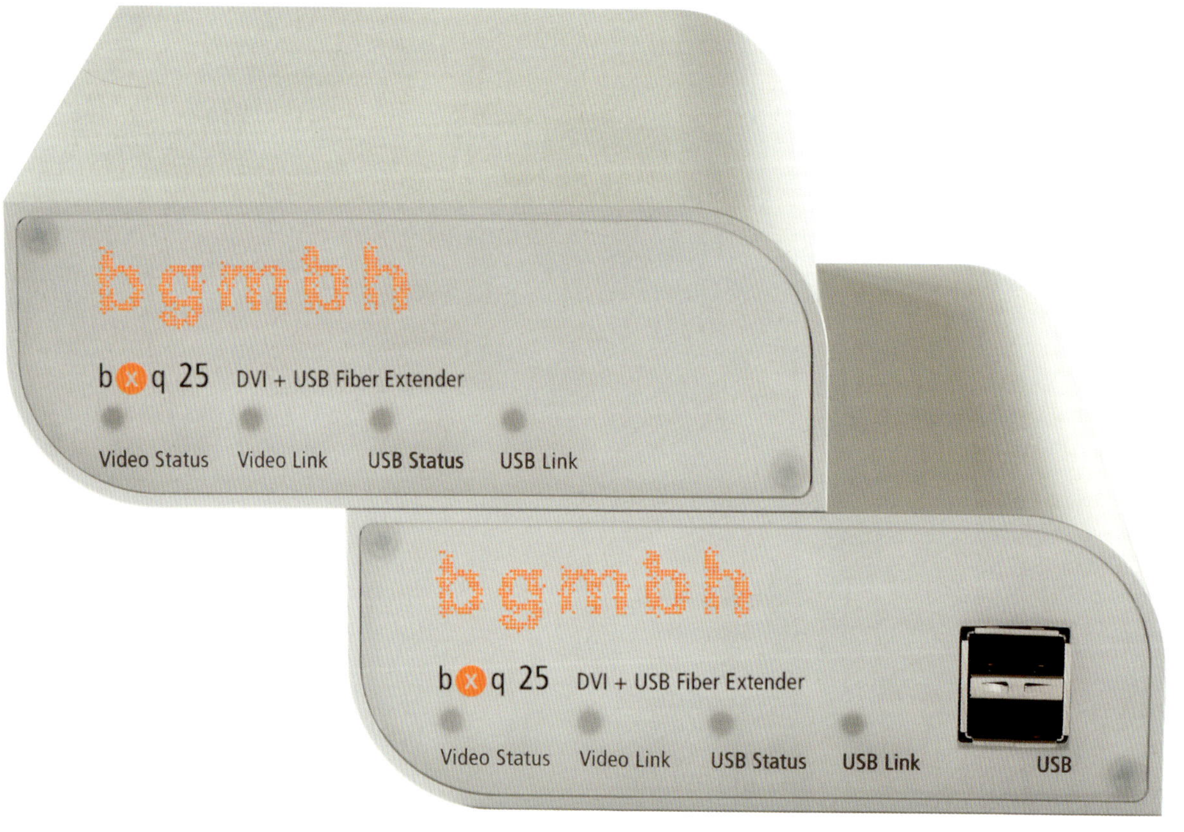

bxq-25

Glasfaser-Bildüberträger
Glass fibre image transmitter

Hersteller / Manufacturer
bgmbh Medien + Datenbanken
D-71636 Ludwigsburg
—
Design / Designer
Schnabel, Schneider Industrial Design
D-73614 Schorndorf
—
Vertrieb / Distributor
bgmbh Medien + Datenbanken
D-71636 Ludwigsburg
—

Der Glasfaser-Bildüberträger »bxq-25« verlängert Videosignale und USB-Schnittstellen über Glasfaserkabel und überbrückt so spielend große Entfernungen, die Multimode-Ausführung bis zu 400 Metern und die Singlemode-Ausführung bis zu 10 Kilometern.

Er ersetzt energieintensive, störende, lokale Komplettrechner durch konzentrierte, energiesparende, kleine und geräuschlose Einheiten. »bxq-25« überträgt perfekt analoge und digitale Videosignale. Anders als bei der Funkübertragung bekommt niemand Zugriff auf die Daten. Deshalb setzen selbst Geheimdienste auf die Glasfaserübertragungstechnik. Durch eine klare Form, selbsterklärende Gestaltung und eine intuitiv logische Gliederung setzt der Glasfaser-Bildüberträger konsequent Akzente in Design und Funktionalität.

The »bxq-25« glass fibre image transmitter »extends« video signals and USB interfaces via glass fibre cables, thus easily bridging long distances—the multimode version up to 400 metres and the single-mode version up to 10 kilometres.

It replaces energy-intensive local computer systems, which are a nuisance, with small, silent, concentrated and energy-saving units. »bxq-25« transmits both analog and digital video signals perfectly. In contrast to transmission by radio, hacking is impossible, which is why even secret services rely on glass-fibre transmission technology. With its clear form, self-explanatory design, and intuitively logical arrangement, the glass fibre image transmitter makes its mark in both design and functionality.

DaimlerChrysler
360
360 GRAD · Umweltbericht 2005
Alles, was uns bewegt

Cocktails für Connaisseure
Nature's Finest

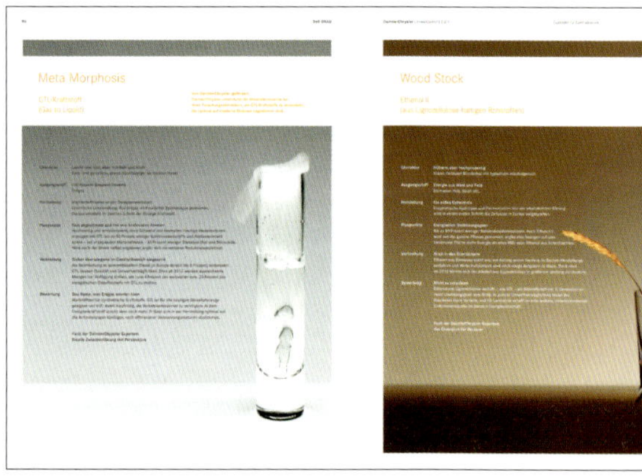
Meta Morphosis
Wood Stock

Flower Power
Deepest Dreams

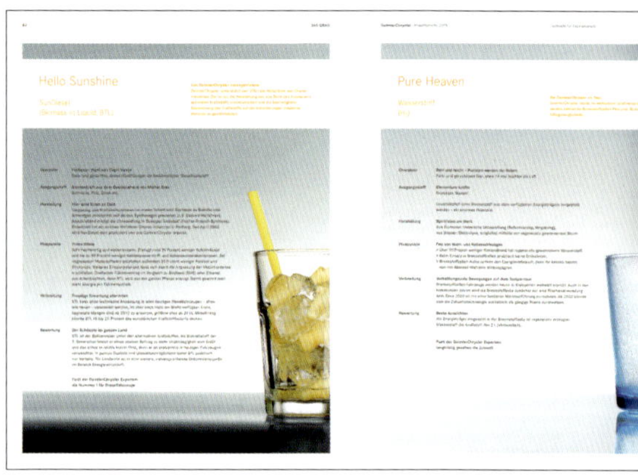
Hello Sunshine
Pure Heaven

360 Grad

DaimlerChrysler Umweltbericht 2005

DaimlerChrysler Environmental Report 2005

Hersteller / Manufacturer
DaimlerChrysler AG
D-70546 Stuttgart
–
Design / Designer
Design hoch drei GmbH & Co. KG
D-70376 Stuttgart
–

Der Umweltbericht 2005 von DaimlerChrysler mit dem Untertitel »Alles, was uns bewegt« kreist um die großen Themen Energie und Emotion. Als Magazin konzipiert, unterhält das Unternehmensmedium seine Leser mit spannenden Geschichten und Reportagen rund um das weltweite Umweltengagement des Automobilkonzerns.

Mit aufregenden Fotostrecken, feinsten Illustrationen und konsequentem Layout werden die oftmals technisch eingegrenzten Themen Umwelt und Energieeinsparung spannend und emotional präsentiert.

The DaimlerChrysler Environmental Report 2005, with the subtitle »The Forces That Move Us«, revolves around the central themes of energy and emotion. A magazine-style corporate medium, it entertains its readers with thrilling stories and reports on various aspects of the group's global environmental commitment.

The environment and energy economy, often only seen from a technological standpoint, are presented in a stimulating, emotional way with exciting photo sequences, superb illustrations and consistent layout.

Jury
Sich mit diesem Thema zu befassen und es dann zum Wettbewerb einzureichen, halten wir für sehr gut. Eine zeitgemäße Darstellung. Als eine angemessene Lösung im Kontext des Wettbewerbs zu bewerten, der sich eigentlich dem Thema Industriedesign widmet.
–
Judges panel
We regard it as a very good idea to work on this topic and then to enter it in a competition. An up-to-date presentation. An appropriate solution in the context of the competition, which is indeed devoted to industrial design.
–

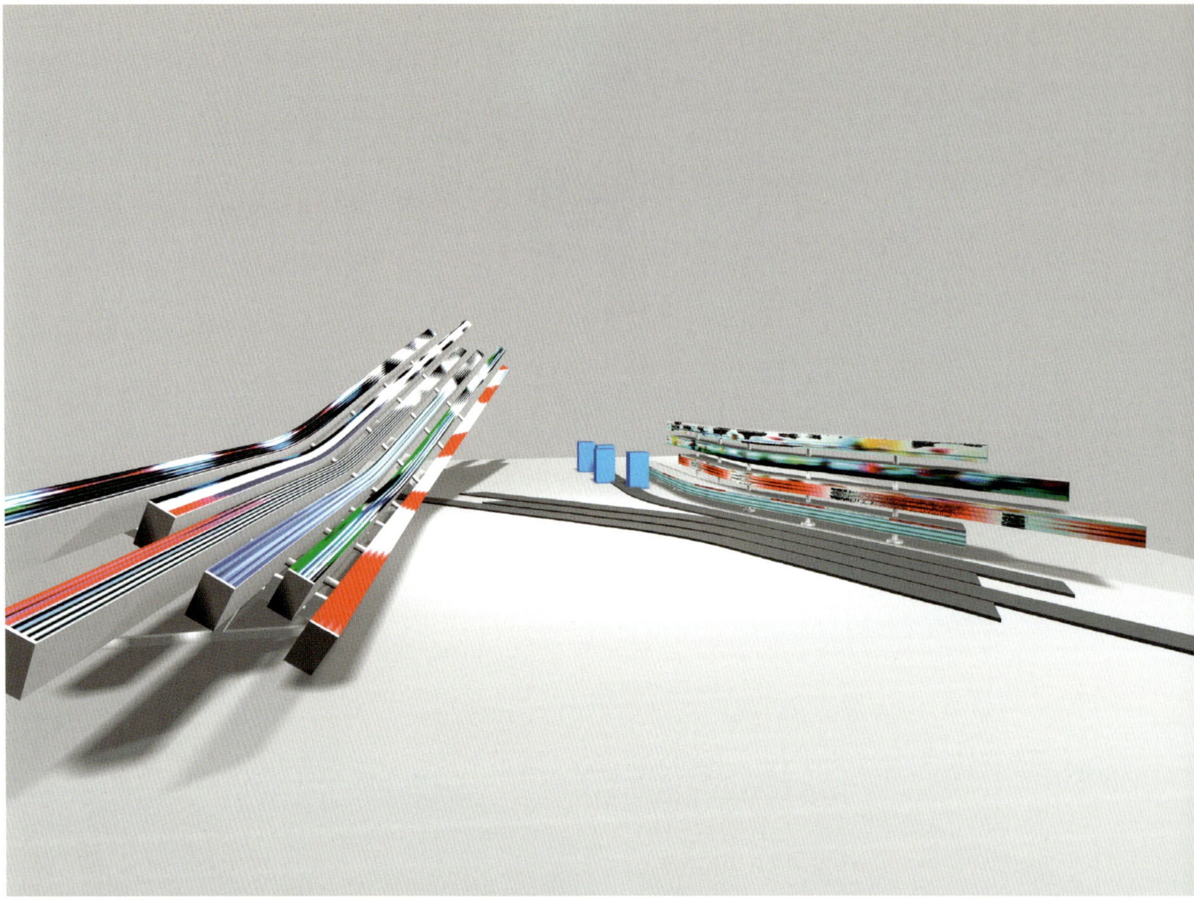

Kimi
Mobil 1
Mercedes-Benz
SIEMENS
SCHUCO
LOCTITE
SONAX

Faszination Motorsport

Wanderausstellung für Kundencenter von Mercedes-Benz

Travelling exhibition for Mercedes-Benz customer centres

Hersteller / Manufacturer
DaimlerChrysler AG
D-70546 Stuttgart
—
Design / Designer
Design hoch drei GmbH & Co. KG
D-70376 Stuttgart
—

Die Wanderausstellung »Faszination Motorsport« wurde für Kundencenter von Mercedes-Benz konzipiert. Sie lädt die Besucher ein, sich in einer atmosphärisch dichten Inszenierung voller Energie und Emotionen vom Motorsport faszinieren zu lassen.

Herzstück des Auftritts bildet eine bumerangförmige, rasant nach oben strebende Steilkurve. Die eindrucksvolle Form ist belegt mit Motiven rund um das Rennfeld. Die dynamische Optik fängt den magischen Moment ein, wo das Verfolgen des Rennautos den Hintergrund zu einem abstrakten Formen- und Farbenspiel verschwimmen lässt.

The travelling exhibition »Fascination Motor Sport« was developed for Mercedes-Benz Customer Centres. It invites visitors to enjoy the fascination of motor sport in an atmosphere-packed production full of energy and emotion.

The centrepiece of the performance is a steep boomerang-shaped curve, forging rapidly upwards. This impressive shape is occupied by subjects associated with the racetrack. The dynamic optics capture the magic moment when following the racing cars causes the background to blur into an abstract interplay of shapes and colours.

Jury
Sehr dynamischer Einsatz von Raum und Geometrie, um einen Eindruck von Geschwindigkeit, Energie und die historische Entwicklung von Motorsport zu vermitteln.
—
Judges panel
Very dynamic use of space and geometry to give the impression of speed, energy, and the history of motor sport.
—

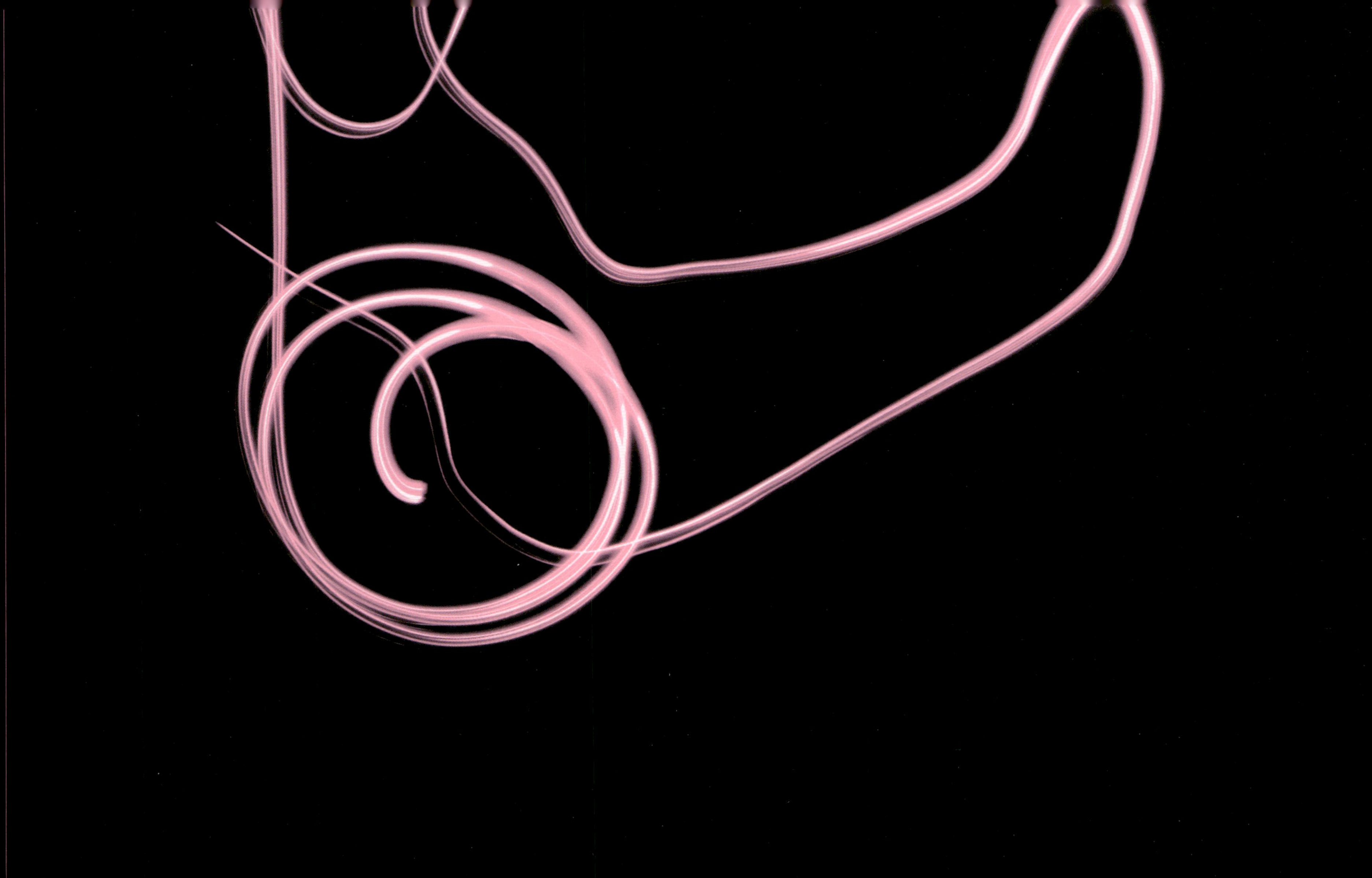

Jury

Sehr hilfreiches Produkt für die große Zielgruppe der älteren Menschen. Die Funktion des kleinen Wendekreises und des Seitwärtsfahrens überzeugt. Sehr gutes Konzept, gutes Produkt für die Zukunft.

—

Judges panel

Very helpful product for a large target group—older people. The small turning circle and the ability to move sideways are impressive features. Very good concept. A good product for the future.

—

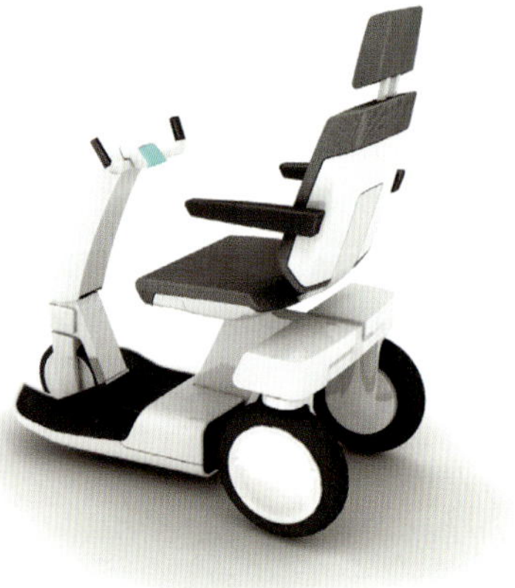

I10

Scooter

Design / Designer
Industrial PDD GmbH & Co. KG
D-70173 Stuttgart
—

Herkömmliche Scooter degradieren meist aufgrund ihrer Gestaltung Nutzer – ältere Menschen, die zunehmend mehr Ansprüche in Bezug auf besseres Lebensgefühl, Unabhängigkeit und Mobilität stellen – zu »Gehandicapten«. Wegen fehlender Wendigkeit oder zu großer Abmessungen können sie nur im Outdoorbereich eingesetzt werden. So müssen sie zum Beispiel vor dem Supermarkt geparkt werden, da sie nicht durch die engen Gänge bewegt werden können.

Der »I10«, der im Rahmen eines Forschungsprojekts entwickelt wurde, lässt sich dagegen durch seine kleinen Abmessungen und seine enorme Wendigkeit nicht nur auf Straßen, sondern auch im Supermarkt und selbst in der eigenen Wohnung einsetzen. Dank einer neuentwickelten patentierten Technologie, bei der die durch Radnabenmotoren angetriebenen Hinterräder über zwei Stellmotoren in einem Winkelbereich von 0 bis 90 Grad nach links und rechts verstellt werden können, besitzt der »I10«, eine einzigartige Wendigkeit. Das macht es möglich, dass er sich sowohl auf der Stelle im Kreis drehen als auch nach links oder rechts parallel fahren kann.

Durch modularen Aufbau werden drei Produktvarianten angeboten, wodurch auch der Scooter-Standardmarkt bedient wird. Abgerundet wird das Angebot durch Accessoires, wie eine Taschen- oder Golfbaghalterung. Dadurch kann der »I10« auch beim Golfsport, Einkaufen oder im Flughafencenter eingesetzt werden.

It is usually due to their design that conventional scooters degrade some users—older individuals who attach increasingly more importance to better quality of life, independence and mobility—to »handicapped« status. Because of their lack of manoeuvrability or their size these scooters can only be used outdoors. For example, they have to be parked outside the supermarket because the aisles between the shelves are too narrow for them.

The »I10«, developed as part of a research project, reflects a new approach. Thanks to its small size and remarkable manoeuvrability it can be used not only on the road but also in supermarkets and even in the owner's home. The unique manoeuvrability of the »I10« is the result of a new, patented technology: by means of two servo motors the angle of the rear wheels, which are driven by hub motors, can be adjusted by up to 90 degrees to the left or right. This enables the scooter not only to rotate on the spot but also to move to the left or right with the wheels parallel to one another.

Thanks to modular construction three different models are available, including one for the standard scooter market. Accessories such as a shopping bag or golf bag holder round off the product range, so that the »I10« can also be used on the golf course, for shopping or at airports.

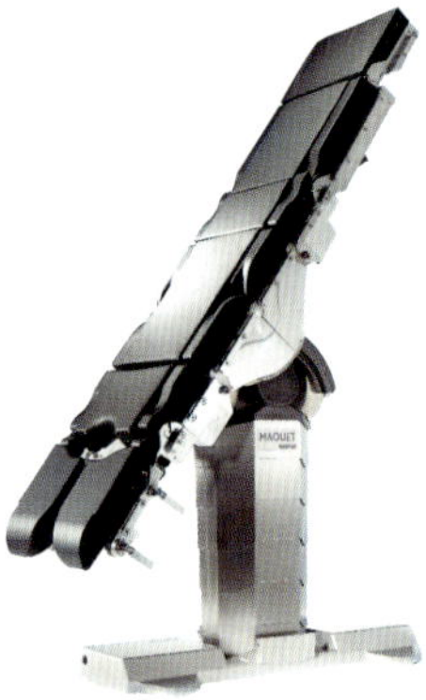

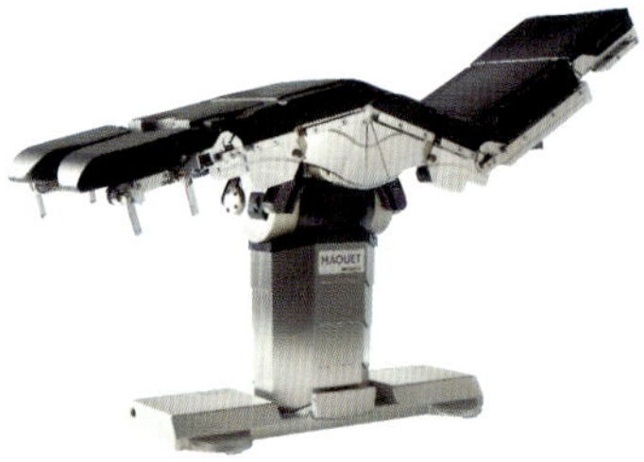

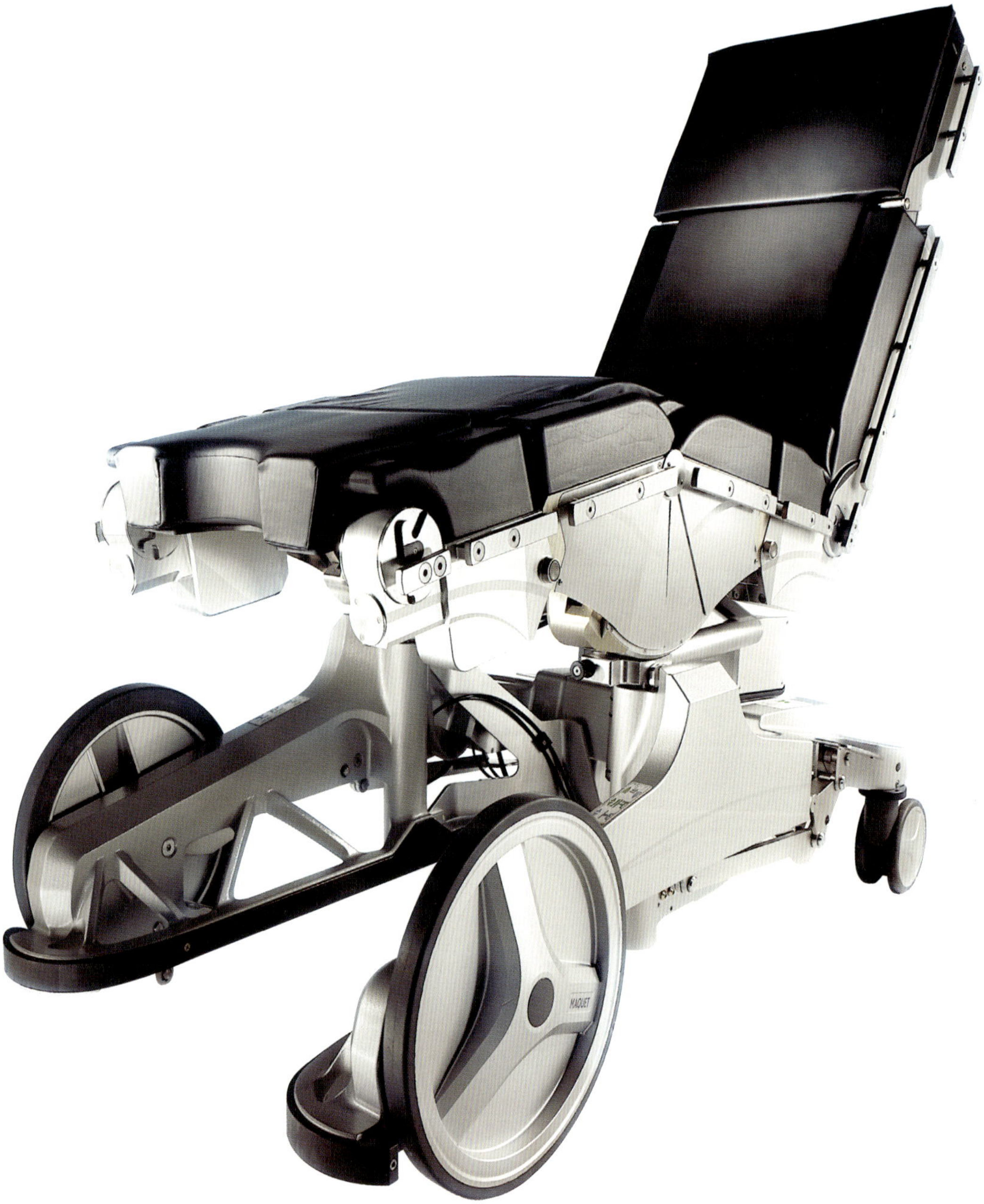

Magnus

OP-Tischsystem
Operating table system

Hersteller / Manufacturer
Maquet GmbH
D-76437 Rastatt
—
Design / Designer
Design Tech
Jürgen R. Schmid
D-72119 Ammerbuch
—
Vertrieb / Distributor
Maquet
Vertrieb & Service Deutschland GmbH
D-76437 Rastatt
—

Das OP-Tischsystem, das sich auch im Design deutlich von konventionellen Systemen absetzt, erlaubt erstmalig, Patienten ohne Umbetten, sitzend oder liegend, von der Station bis in den OP zu fahren. Das schont nicht nur Patienten, sondern spart auch Zeit und Kosten. Dank einem Spezialfahrwerk des Transporters lässt sich »Magnus« bequem von einer einzigen Pflegekraft, statt wie bisher zwei, manövrieren. Große Räder vorne und zwei Lenkrollen hinten bieten auch bei Unebenheiten, wie Aufzugsschwellen, hohen Fahrkomfort.

Im OP angekommen wird die Lagerfläche sicher und komfortabel vom Transporter auf die OP-Tischsäule übernommen. Bisher ungekannte, extreme Verstellbereiche der Lagerfläche machen »Magnus« zum modernsten und leistungsfähigsten OP-Tisch- und Transportsystem. Das erlaubt nicht nur eine entspannte ergonomisch gute Arbeitshaltung für das OP-Team sondern auch eine deutlich bessere Patientenlagerung, die sämtliche chirurgische Eingriffe ermöglicht.

Der modulare Aufbau bietet höchste Flexibilität für verschiedene chirurgische Fachdisziplinen beziehungsweise für unterschiedliche Körpergrößen des Patienten. Das neu entwickelte kompakte Handbediengerät besitzt erstmals ein beleuchtetes Display und selbsterklärende Symboltasten, die eine schnelle und intuitive Bedienung erlauben.

With this operating table system patients can for the first time ever be moved directly from the ward to the operating theatre, seated or supine, without having to change beds. This not only reduces patient stress, it also saves time and money. The design of the system is vastly different from that of conventional systems. Thanks to a special transporter chassis, »Magnus« can be manoeuvred easily by just one nurse. Large wheels at the front and two castors at the rear ensure high comfort even on uneven surfaces such as lift entrances.

In the operating theatre, the support surface is transferred safely and conveniently from the transporter to the operating table column. Thanks to the extreme, previously impossible adjustment range of the support surface, »Magnus« is currently the most up-to-date and efficient operating table and patient transport system. It not only allows the surgical team to adopt relaxed, ergonomically correct postures, but also permits far better patient positioning, making all types of surgical interventions possible.

The modular construction offers maximum versatility for various surgical disciplines and for patients of different sizes. The new, compact manual control unit now has an illuminated display and self-explanatory, symbol-marked buttons for fast, intuitive operation.

Jury
Alle Produkte, die Schnittstellen zwischen OP, Intensivstation und auch noch Diagnosegeräten optimieren, gestalten den work flow im Krankenhaus effizienter. Dabei hilft das neue Maquetprodukt erheblich, weil die Patienten weniger umgebettet werden müssen und in flexibleren Körperpositionen transportiert und sogar operiert werden können.
—

Judges panel
Every product that optimizes the interfaces between operating theatre, intensive care unit and diagnostic devices makes the workflow in hospitals more efficient. The new product by Maquet makes a major contribution in this respect, because the patients have to be rebedded less often, and can be transported or even operated on in a wider variety of anatomical positions.
—

Multifunktionsdusche / Multifunctional shower

Multifunktionsdusche
Multifunctional shower

Hersteller / Manufacturer
Duravit AG
D-78132 Hornberg
—
Design / Designer
Schmiddem
D-10777 Berlin
—
Vertrieb / Distributor
Duravit AG
D-78132 Hornberg
—

Die Multifunktionsdusche ist nicht nur zum Duschen, sondern auch zum Dampfbaden und Relaxen. Zehn Massagedüsen, Kopf-, Hand- und Schwallbrause, Kneippschlauch oder die Funktionsmodule für Duft, Farblicht oder Sound sorgen für Wellness im Privathaushalt—und das stehend oder liegend, allein oder zu zweit.

Die rahmenlose Echtglaskabine, wahlweise in klarem oder blau eingefärbtem Einscheibensicherheitsglas, passt sich optimal in die vorhandene Badarchitektur ein, wirkt von außen optisch leicht und dezent, von innen bietet sie reichlich Bewegungsfreiheit und Weitblick.

Das Wasser zirkuliert in einem geschlossenen Kreislauf, was zu einem minimalen Wasserverbrauch führt. Mit nur 40 Liter Wasser und Düsen mit großem Durchmesser für den Wasserauslass erzielt die Multifunktionsdusche maximale Effekte für kraftvolle Körpermassagen. Mit minimalem Energieverbrauch (18 Watt) und ganz ohne Chemie sorgt die UV-Desinfektionsanlage für bestmögliche Hygiene. Eine neue Art der Dampferzeugung gewährleistet die optimale Ausbeute der eingesetzten Energie. Das Farblicht-Modul ist mit fünf energiesparenden LED-Strahlern mit insgesamt 20 Watt ausgestattet.

The multifunctional shower is designed not only for showering, but also for taking saunas and relaxing. Ten massage nozzles, head, hand and torrent shower, Kneipp hose, and function modules for fragrance, coloured light and sound ensure wellness at home, whether standing or lying down, alone or with a partner.

The frameless cubicle, made of clear or blue-tinted single-pane safety glass, integrates ideally into existing bathrooms. From the exterior it makes a light and discreet impression, while from the interior it offers ample freedom of movement and vision.

The water circulates in a closed circuit, which minimizes water consumption. The multifunctional shower achieves its maximum effect—for powerful body massage—with just 40 litres of water and large-diameter water outlet nozzles. With minimal energy consumption (18 watts) and no chemicals at all, the UV disinfection system ensures the best possible level of hygiene. A new method of generating steam guarantees optimal yield from the energy. The colour light module is equipped with five energy-saving LED spots, with a total rating of 20 watts.

Jury
Besonders interessant für ältere Menschen, denen es schwerfällt in eine ganz normale Badewanne einzusteigen. Gut die Energieeinsparung: Beim Duschen braucht man viel weniger Heizenergie für Warmwasser als beim Baden. Und nicht zu vergessen der Wellness-Aspekt zur Revitalisierung der persönlichen Energie.
—

Judges panel
An interesting option in particular for elderly people who have difficulty getting into an ordinary bathtub. Good energy economy. Much less energy is needed to heat water for showering than for taking a bath. And the wellness effect—revitalization—is not to be neglected.
—

Jury

Ein Gerät mit einer innovativen Technologie, die das Wassererwärmen energieeffizienter macht, also insofern genau passend zum Wettbewerbsthema »Focus Energy«. Überzeugend vor allem der intelligente Einsatz von Technik.

—

Judges panel

A device with innovative technology which makes water heating more energy-efficient, and thus highly relevant to the competition theme »Focus Energy«. Above all, the intelligent application of technology is impressive.

—

DSX Servotronic MPS®

Durchlauferhitzer
Instantaneous water heater

Hersteller / Manufacturer
Clage GmbH
D-21337 Lüneburg
—
Design / Designer
Design Planet
José Delhaes
D-63165 Mühlheim am Main
—
Vertrieb / Distributor
Clage GmbH
D-21337 Lüneburg
—

»Servotronic« bietet Top-Technik in einem raumsparenden Design, das sich harmonisch in jede Badumgebung einfügt. Das Gerät ist einfach zu bedienen. Eine der vier Anwendungstasten drücken, und das Wasser hat die richtige Temperatur zum Händewaschen (35°C), Duschen (38°C), Baden (42°C) oder Geschirrspülen (48°C). Die Temperatur lässt sich aber auch direkt zwischen 20°C und 60°C auswählen. Das Farbdisplay signalisiert rot für heiß bis blau für kalt.

Das Gerät mit dynamischer Durchflussregelung sorgt für konstant gradgenaue Auslauftemperaturen und maximale Wirtschaftlichkeit. Es erwärmt das Wasser nur dann, wenn es tatsächlich gebraucht wird, exakt auf die gewählte Nutztemperatur, und zwar direkt an der Zapfstelle. Bei zentralen Systemen dagegen entstehen Energieverluste, wenn heißes Wasser verteilt und erst an der Armatur nach langen Leitungswegen auf die tatsächliche Nutztemperatur heruntergemischt werden muß. Zusätzlich wird wertvolles Trinkwasser gespart, das bei zentralen Systemen ungenutzt in den Abfluss läuft.

Der Energiemonitor macht die Sparsamkeit des Gerätes deutlich. Die Textanzeige hält bei Bedarf zahlreiche Betriebs- und Serviceinformationen bereit. So lassen sich zum Beispiel Kaltwassertemperatur, Durchflussmenge und Stromverbrauch ablesen.

»Servotronic« offers leading-edge technology in a space-saving design that fits harmoniously into any bathroom. The device is easy to operate. Pressing one of the four buttons produces a flow of water at the right temperature for washing hands (35°C), showering (38°C), taking a bath (42°C) or washing dishes (48°C). The temperature can also be set directly, between 20°C and 60°C. The colour display indicates red for hot to blue for cold.

With dynamic flow control, the heater ensures constant outflow temperatures and maximum economy. It heats the water only when it is actually needed, exactly to the required temperature and directly at the tapping point. By contrast, in central systems energy is wasted when hot water is distributed through long pipes and mixed to the required temperature at the tap. In addition, the heater saves precious drinking water, which in central systems flows down the drain unused.

The energy monitor makes it clear how economical the heater is. The text display can show a wide range of operating and service information as required, for example cold water temperature, volumetric flow rate and power consumption.

BOSCH

BOSCH

LiftMatic

Wandbackofen
Wall-mounted oven

Hersteller / Manufacturer
Robert Bosch Hausgeräte GmbH
D-81739 München
—
Design / Designer
Werksdesign / In-house design
Roland Vetter
—
Vertrieb / Distributor
Robert Bosch Hausgeräte GmbH
D-81739 München
—

Mit dem Wandbackofen »LiftMatic« eröffnen sich nahezu unbegrenzte Möglichkeiten der Küchengestaltung. Ob zwischen Oberschränken eingebaut oder frei an der Wand, der attraktive »LiftMatic« wird sofort zum vielbeachteten Designmittelpunkt in der Küche.

Hinzu kommt ein innovatives Technikkonzept mit neuen Dimensionen an Sicherheit und Komfort. Mit der integrierten Touch Control-Blende werden alle Funktionen komfortabel und einfach angesteuert. Auf Tastendruck fährt bei LiftMatic der Backofen-Innenraum auf und ab und stoppt automatisch, wenn sich ein Hindernis unter dem Gerät befindet. Über die optimale Innenraum-Ausleuchtung hat man jedes Detail im Blick.

Teller können mit einer eigenen Funktion vorgewärmt werden. Darüber hinaus können Speisen im geöffneten wie geschlossenen Zustand auf dem Glaskeramik-Boden warmgehalten werden.

The »Liftmatic« wall-mounted oven opens up almost unlimited possibilities for kitchen design. Whether it is placed between two wall cupboards or mounted on its own, the attractive »LiftMatic« instantly stands out as a striking design feature in the kitchen.

But that is not all. Its innovative technology also sets new standards in safety and convenience. Its integrated touch control panel allows all functions to be controlled conveniently and simply. At the touch of the button, »Liftmatic« raises and lowers the inner shell of the oven like a dumb waiter, and stops automatically if it encounters an obstacle below the appliance. The excellent interior lighting leaves no corner unlit.

The oven is equipped with a separate plate-warming function. Moreover, food can be kept hot on the ceramic glass base, irrespective of whether the oven is open or closed.

Jury
Das Konzept eines herunterfahrbaren Ofens ist in kleinen Küchen sehr sinnvoll. Einfache und intuitive Bedienung, von beiden Seiten zugänglich. Zurückhaltende Gestaltung ist gerade im Küchenbereich sehr passend. Leicht zu bedienen, gute ergonomische Details. Die Höhe ist für Öfen ideal, weil man gut sieht, wie der Braten wird.
—
Judges panel
The idea of an oven floor that can be raised or lowered makes a lot of sense for small kitchens. Simple, intuitive operation, and accessible from both sides. Its unobtrusive styling is very appropriate for the kitchen environment. Easy to operate, good ergonomic details. Height like this is ideal for an oven, because you can easily keep an eye on how the Sunday joint is doing.
—

Hausgeräte / Silber
Household appliances / Silver

TK 68 E 570 + HW 140 560
Kaffeevollautomat / Fully automatic coffee maker

TK 68 E 570 + HW 140 560

Kaffeevollautomat mit Wärmeschublade

Fully automatic coffee maker with hot cupboard

Hersteller / Manufacturer
Siemens Electrogeräte GmbH
D-81739 München
—
Design / Designer
Werksdesign / In-house design
Gerd G. Wilsdorf,
Frank Rieser, Heiko Thielen
—

Die voll integrierbare Espressomaschine passt in das kompakte Einbaumaß von 45 x 60 Zentimetern. Sie ist mit höhenverstellbarem Kaffeeauslauf und Dampfdüse ausgestattet und leicht zu bedienen. Über Klartextanzeige wird der Benutzer einfach durch die Funktionen geführt.

Die kompakte Funktionseinheit aus Einbau-Kaffeevollautomat und Wärmeschublade hat zusammen ein Einbaumaß von 60 Zentimetern in der Höhe. In der Wärmeschublade können Espressotassen vorgewärmt werden. Kaffeevollautomat und Wärmeschublade nehmen die klare elegante Formensprache der Siemensgeräte auf. Glasblende mit Edelstahlbedienzone und funktionale puristische Griffgestaltung zeichnen sie aus.

This espresso machine can be integrated perfectly into a slot measuring 45 by 60 centimetres. It is equipped with a height-adjustable coffee dispensing spout and steam arm, and is easy to operate. The plain text display provides an easy-to-follow guide for the user.

In combination with a hot cupboard, the coffee maker forms a compact unit measuring just 60 centimetres in height. The hot cupboard is ideal for warming espresso cups. With its glass panel, a stainless steel operating zone and the functional, purist design of the knobs, the design of the fully automatic coffee maker and hot cupboard reflects the clear formal expression of Siemens appliances.

Jury
Integrierte Espressomaschine, leicht zugänglich, mit integriertem Tassenwärmer, formal schlicht. Schönes Design, freundlich im Gebrauch, sehr gute Details, leicht zu reinigen.
—
Judges panel
Integrated espresso machine, easily accessible, with an integrated cup warmer, formally unassuming. Elegant design, user-friendly, excellent details, easy to clean.
—

WMF 1000
Kaffeemaschine / Coffee maker

WMF 1000

Kaffeemaschine
Coffee maker

Hersteller / Manufacturer
WMF Württembergische Metallwarenfabrik AG
D-73312 Geislingen
—
Design / Designer
Metz & Kindler Produktdesign
D-64293 Darmstadt
—
Vertrieb / Distributor
WMF Württembergische Metallwarenfabrik AG
D-73312 Geislingen
—

Der Kaffeeautomat »WMF 1000« in matt-glänzendem Cromargan® Gehäuse ist eine Profimaschine für zu Hause. Er vereint perfekte Technik, innovative Funktionen, hochwertige Materialität und außergewöhnliches, funktionales Design.

Er bietet die Originaltechnologie, die sich in Gastronomie-Maschinen bewährt hat. So arbeitet der WMF Profi-Brüher mit dem »WMF Vorbrüh-Prinzip« in drei Schritten: Das Kaffeemehl wird vor dem Brühvorgang mit heißem Wasser benetzt. Die Kaffeepulverkörner quellen auf, so dass sich das Kaffee-Aroma entwickeln kann. Erst dann kann sich das volle Aroma unter hohem Druck im 86° heißen Brühwasser entfalten.

»WMF 1000« ist einfach mit einem zentral angeordneten Multifunktionsschalter, zugleich ein augenfälliges Gestaltungselement, zu bedienen. Auf Knopfdruck bereitet er zwölf verschiedene internationale Spezialitäten von Café Crème bis Latte Macchiato. Zudem lässt sich die Getränkemenge an jede Tasse individuell anpassen.

In its matt-polished Cromargan® housing, the »WMF 1000« automatic coffee maker is a professional-standard appliance for the home. It combines perfect technology, innovative functions, high-quality materials and an unusual, functional design.

Users can benefit from the same technology that has proved its worth in coffee makers for the catering trade. For example, the professional WMF coffee maker works in three stages, according to the »WMF pre-brew principle«: before the actual coffee brewing process, the ground coffee is wetted with hot water. The grains of ground coffee swell, and this allows the coffee aroma to develop. The coffee's full aroma then develops when water is added under high pressure at a temperature of 86°C.

With its centrally located multifunction switch, which is also a striking design element, the »WMF 1000« is easy to operate. At the touch of a button it can prepare twelve different international coffee specialities, from café crème to latte macchiato. In addition, the amount of coffee can be adjusted individually to any cup size.

Jury
Die Maschine bringt Professionalität in die private Küche. Sie macht nicht nur professionelles Kaffeebrühen aus frisch gemahlenen Bohnen für zu Hause möglich, sondern hat auch durch ihre Klarheit ein professionelles Gesicht. Durch einfache Benutzerführung ist es jedermann möglich, das Gerät schnell zu verstehen. Ideale Maschine, um auch unangemeldete Besucher im Büro gut bedienen zu können.
—
Judges panel
This machine lends the private kitchen an air of professionalism. It not only allows you to make coffee professionally from freshly ground beans in the home, but also looks professional with its clear lines. The simple user controls make it possible for anyone to understand the appliance quickly. Ideal machine if you want to be able to serve unannounced guests in the office, too.
—

250°C
12:15
BOSCH

HLN 45 44 50

Elektro-Standherd
Free-standing electric cooker

Hersteller / Manufacturer
Robert Bosch Hausgeräte GmbH
D-81739 München
—
Design / Designer
Werksdesign / In-house design
Roland Vetter und
Eisele Kuberg Design
Frank Eisele
D-89231 Neu-Ulm
—
Vertrieb / Distributor
Robert Bosch Hausgeräte GmbH
D-81739 München
—

Der Elektro-Standherd zeichnet sich durch hohen Bedienkomfort aus. Das Glaskeramik-Kochfeld verfügt über eine Zweikreis-Kochzone und eine zuschaltbare Bräterzone. Die Restwärme wird für alle vier Kochzonen angezeigt.

Der versenkbare Knebel verhindert ein versehentliches Einschalten. Der Backofen ist mit einem komfortablen Backwagen ausgestattet. Programme wie automatische Schnellaufheizung, Klein- und Großflächengrill, Heißluftgrill und Auftau-Automatik gehören zur Grundausstattung des Elektro-Standherds.

The distinctive feature of this free-standing electric cooker is its great ease of use. The ceramic glass hob has one dual-circuit cooking zone and an additional frying zone. Residual heat is displayed for all four cooking zones.

The retractable control knob stops the machine from being switched on unintentionally. The oven is equipped with a convenient oven carriage. Programs such as automatic fast preheat, small and large grill, hot-air grill and automatic thawing are standard features in this electric free-standing cooker.

Jury
Einfacher klarer Standherd im Edelstahldesign. Hochwertiges Erscheinungsbild, auch für kleine Mietwohnungen geeignet. Bringt ein bißchen Profiküche in den Privathaushalt. Klares leicht zu verstehendes Bedienfeld, einfach sauber zu halten. Kann sowohl frei stehen als auch in eine bestehende Küchenzeile integriert werden, weil nicht nur die Front sondern auch die Seite aus Edelstahl ist. Schick, modern, zeitgemäß.
—
Judges panel
Simple, clear free-standing cooker in stainless steel design. High-quality appearance, also suitable for small rented flats. Adds a touch of the professional kitchen to the private household. Easily understandable operating panel, easy to keep clean. Can either stand on its own or be integrated into an existing fitted kitchen, since both the front and the side panels are made of stainless steel. Chic, modern, contemporary.
—

BOSCH

Piano à filo PCL 795 FEU

Gaskochstelle
Gas hob

Hersteller / Manufacturer
Robert Bosch Hausgeräte GmbH
D-81739 München
—
Design / Designer
Werksdesign / In-house design
Alexander Marsch
—
Vertrieb / Distributor
Robert Bosch Hausgeräte GmbH
D-81739 München
—

»Piano à filo« ist ein Novum unter den Gaskochmulden. Die Gaskochstelle wird plan zur Arbeitsfläche eingebaut. Das führt zu einem optischen Blickfang mit praktischen Vorteilen, beispielsweise ist sie ideal zum Reinigen.

Die fünfflammige Edelstahl-Gaskochstelle besitzt Einzeltopfträger: Neben zwei Normal- und einem Sparbrenner ist »Piano à filo« auch mit einem Wok- und einem Großbrenner ausgestattet. Die Brenner mit Knebelbedienung sind sowohl für Erdgas als auch Flüssiggas geeignet. Die Gaskochstelle zeichnet sich durch ihre klare, funktionale, übersichtliche Gestaltung aus.

»Piano à filo« is a new kind of gas hob. The gas hob is fitted flush with the worktop. This makes it both an eye-catcher and has practical advantages, since it is very easy to clean.

The five-burner stainless steel gas hob has individual pan supports: »Piano à filo« is equipped with one wok, one large, two standard and one economy burner. Operated by control knobs, the burners are suitable for both natural and liquid gas. The gas hob stands out by virtue of its clear, functional and uncluttered design.

Jury
Sehr praktischer Gasherd, sehr einfach zu reinigen, ganz klare schlichte Gestaltung, glatte Oberflächen. Also für jede Hausfrau und jeden Hausmann ein Vorteil. Die Bedienelemente sind klar und sauber gestaltet. Sehr gut gemacht, dass fünf Flammen auf kleinstem Raum untergebracht sind. Ein guter Einsatz von Energie.
—
Judges panel
Very practical gas cooker, very easy to clean, very clear, unassuming design, smooth surfaces. The latter feature is an advantage for any housewife or househusband. The operating elements are clearly and neatly designed. Very well done, the way five burners have been fitted into a very small space. A good application of energy.
—

Hausgeräte / Silber
Household appliances / Silver

ER 74 251 ME
Gaskochstelle / Gas hob

ER 74 251 ME

Gaskochstelle
Gas hob

Hersteller / Manufacturer
Siemens Electrogeräte GmbH
D-81739 München
—
Design / Designer
Werksdesign / In-house design
Frank Rieser
—

Die Gaskochstelle ist mit zwei neu entwickelten Wok-Gasbrennern ausgestattet, die eine unterschiedliche Wärmeabstrahlrichtung haben. Dadurch wird den Bedürfnissen nach »Kochen im großen Topf«, im Wok, besonders Rechnung getragen.

Da die Edelstahlfläche kaum verformt ist, lässt sie sich leicht reinigen. Die Primärluft wird durch den hinteren »Spoiler« zugeführt, ein besonders charakteristisches Designmerkmal. Die aus dem Edelstahl geformten »Absteller« sorgen neben ihrer eigentlichen Funktion auch für die Stabilität des Kochfeldes bei thermischer Beanspruchung.

This gas hob is fitted with two newly developed wok burners which give off heat in different directions. This especially takes account of the need to use a wok as a big cooking pot.

As the stainless steel surface is virtually smooth, it can be cleaned easily. Primary air is supplied via the rear »spoiler«, which is an especially distinctive design feature. The surfaces that have been formed from the stainless steel, and serve as a rest for pots, have a further function, since they keep the hob stable when it gets hot.

Jury
Sehr gefälliges Designprodukt, sehr ergonomisch. Ein expressives Produkt für zu Hause, fügt sich gut in die Umgebung von vielen Küchen ein. Und es ist einfach zu reinigen. Gas ist eine gute Energiequelle, ideal zum Kochen.
—
Judges panel
Very pleasing designer product, very ergonomic. An expressive product for the home, harmonizes well with many different kitchen environments. And it is easy to clean. Gas is a good source of energy, ideal for cooking.
—

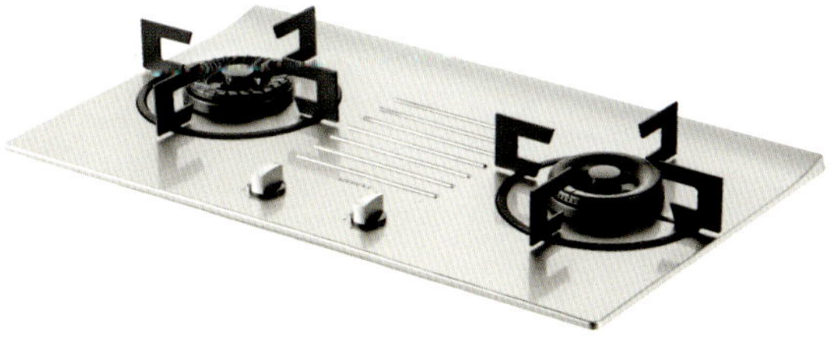

Baureihe W 4000

Waschvollautomat
Fully automatic washing machine

Hersteller / Manufacturer
Miele & Cie. KG
D-33332 Gütersloh
—
Design / Designer
Werksdesign / In-house design
—
Vertrieb / Distributor
Miele & Cie. KG
D-33332 Gütersloh
—

Der Waschvollautomat der hochwertigen Baureihe W 4000 bietet nicht nur bewährte, erstklassige Technik, wie die patentierte Schontrommel für eins bis sechs Kilogramm. Er ist auch mit dem speziellen FU-Antrieb ausgestattet. Dieser ist für besondere Ansprüche entwickelt worden und arbeitet ohne Kohlebürsten. Deshalb ist er verschleißarm, wartungsfrei und noch leiser als die anderen Miele-Motoren.

Eine neue Blendenform, die Pultblende, ermöglicht eine komfortable Bedienung von oben, da die Programme optimal im Blickwinkel liegen. Ein Vor- und Hinabbeugen ist damit nicht mehr nötig. Eine integrierte, wartungsfreie Power-LED leuchtet die Schontrommel perfekt aus.

Das typische Erkennungsmerkmal, der Chromring, glänzt in neuer Form. Seine großzügige Gestaltung harmoniert perfekt mit dem Gehäusedesign. Doch er verleiht dem Gerät nicht nur ein unverkennbares, elegantes Aussehen, sondern hält zudem den härtesten Belastungen stand. Das robuste Material ist besonders schlag-, stoß- und kratzfest.

This fully automatic washing machine from the W 4000 series offers proven first-class technology, such as the patented garment-protecting drum for loads between one and six kilograms. At the same time, however, it is equipped with a special FU drive. This has been developed to meet the most exacting requirements, and works without carbon brushes. This makes it long-wearing, maintenance-free and even quieter than other Miele motors.

A new form of panel, which makes the programme display clearly visible from above, means the washing machine can be operated with ease. There is no longer any need to bend over or down. An integrated, maintenance-free power LED provides excellent lighting for the drum.

Miele's distinguishing feature, the chrome ring, shines in a new form. Its generous styling harmonizes perfectly with the housing design. The ring does not just give the appliance its unmistakable, elegant appearance, however, but can also withstand the toughest conditions. Its robust material is especially shockproof and scratch resistant.

Jury
Waschmaschine mit intelligenten Funktionen, leicht zu nutzen. Die Bienenwabenstruktur der Trommel scheint innovativ zu sein und man hat versucht, mit wenig Materialeinsatz trotzdem hohe Konstruktionssteifigkeit zu erreichen. Diese Trommel hat die Jury überzeugt.
—
Judges panel
Washing machine with intelligent functions, easy to use. The honeycomb structure of the drum appears to be innovative. An attempt has been made to achieve structural rigidity despite the sparing use of materials. This drum impressed the jury.
—

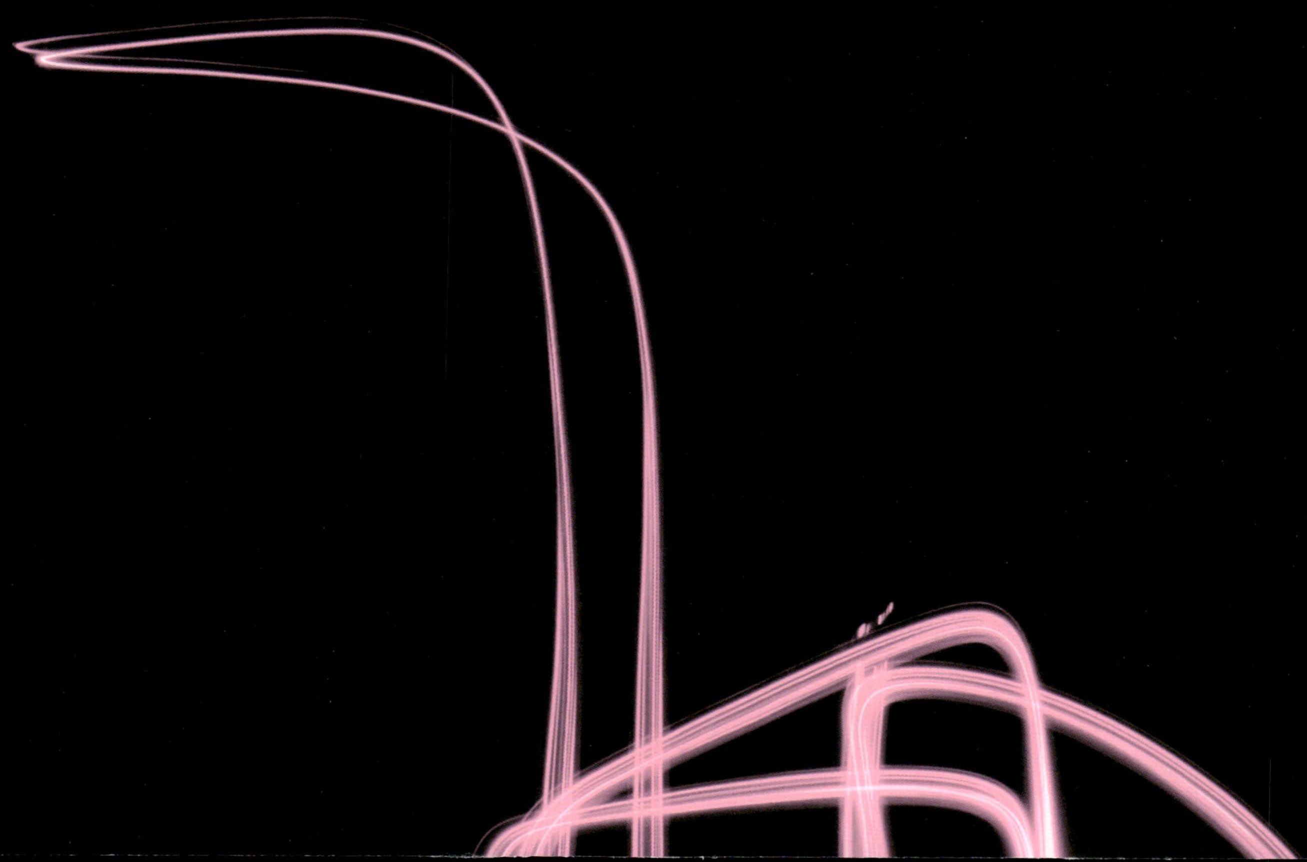

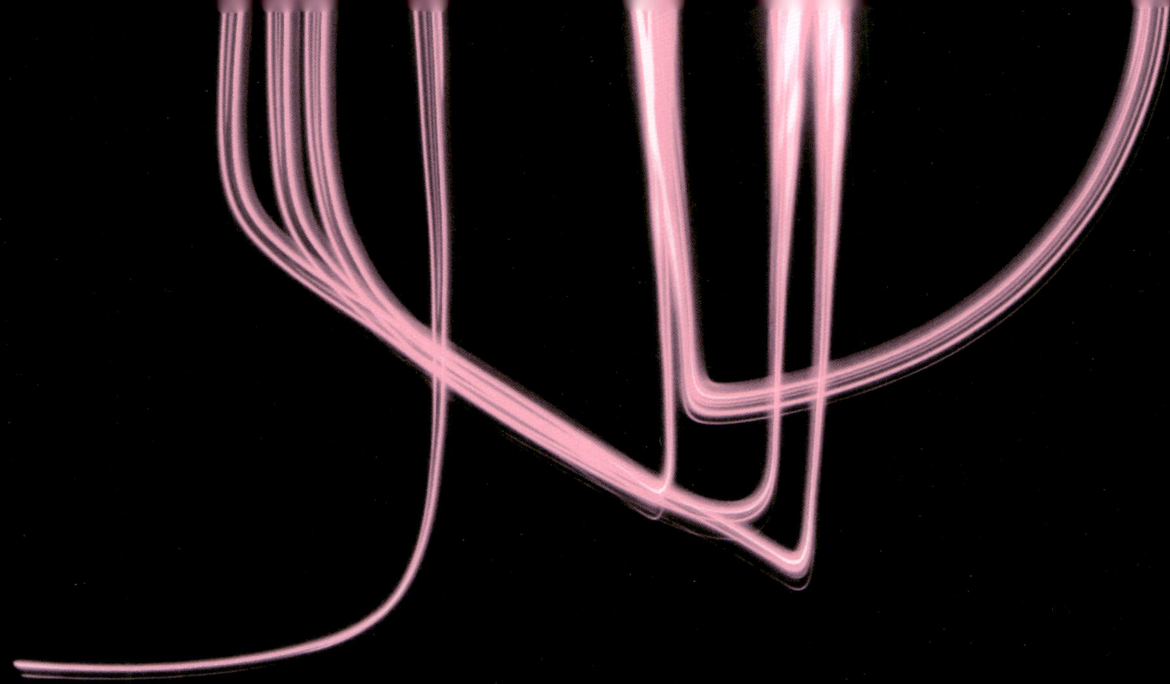

outclass
Tischgrill mit Griff / Tabletop barbecue with handle

outclass
Tischgrill mit Griff / Tabletop barbecue with handle

outclass

Tischgrill mit Griff
Tabletop barbecue with handle

Hersteller / Manufacturer
Eva Denmark A/S
DK-2760 Måløv
—
Design / Designer
Claus Jensen & Henrik Holbæk
Tools®
DK-2400 Kopenhagen NV
—
Vertrieb / Distributor
Eva Denmark Vertriebsleitung Deutschland
Thomas Schmidt
D-56112 Lahnstein
—

Mit diesem Grill auf dem Tisch können alle Personen am Grillvergnügen teilhaben. Die Außenschale besteht aus Porzellan, der Kohlenbehälter und der Rost dagegen aus Edelstahl. Dank eines Stahlbügelgriffs kann der noch kalte Grill vor dem Anzünden der Kohle zum Tisch getragen und auf einem Untersetzer abgestellt werden.

Der Grill funktioniert folgendermaßen: Ein Halter aus Edelstahl wird in die Porzellanschale gesetzt. Darauf kommt ein tellerförmiger Hitzeschutz, der die Hitzestrahlung zur Tischplatte hin vermindert und Beschädigung einer wärmeempfindlichen Unterlage verhindert. Darauf folgen Kohlenbehälter und Rost. Für das Grillvergnügen sind weitaus weniger Kohlen nötig, als bei einem gewöhnlichen Grill.

Nach dem Grillen können Schale, Kohlenbehälter und Rost einfach in der Geschirrspülmaschine gereinigt werden, nur der Hitzeschutz und der Untersetzer müssen von Hand gesäubert werden. Die Porzellanschale besitzt mehrere Funktionen. Wenn sie gerade nicht fürs Grillen benutzt wird, lässt sie sich ohne Einsatz zudem als Salatschüssel verwenden.

With this barbecue on the table, everyone present can share the pleasure of cooking over charcoal. The porcelain outer shell contains a charcoal holder and a grid made of stainless steel. The barbecue has a steel handle that allows it to be carried to the table before lighting the charcoal.

The barbecue works as follows: A stainless steel holder is placed in the porcelain bowl. A plate-like heat shield placed on the holder reduces thermal radiation towards the table and prevents damage to heat-sensitive table-mats etc. Then the charcoal holder and grill are placed in the bowl. The »outclass« burns far less charcoal than a conventional barbecue.

And after the meal the bowl, charcoal holder and grill can be put into the dishwasher. Only the heat shield and the tray have to be cleaned by hand. The porcelain bowl is multifunctional: When it is not being used for a barbecue it can also be used—without the insert—as a salad bowl.

Jury
Netter kleiner Tischgrill für den Balkon, beispielsweise für eine Zielgruppe, die über keinen Garten verfügt, ideal für Single-Haushalte oder Paare. Die Elemente sind leicht auseinanderzunehmen und zu reinigen. Das Material, Porzellan, in diesem Zusammmenhang ist neu und interessant. Vielseitig verwendbar.
—
Judges panel
Nice little tabletop barbecue for the balcony, e.g. for consumers who have no garden. Ideal for one-person households or couples. The components are easy to disassemble and clean. The material—porcelain—is new and interesting in this context. Versatile.
—

Jury

Innovativer Weg, um Messer zu schärfen, der sich von traditionellen Messerschärfern unterscheidet. Auf der Arbeitsfläche allzeit präsentes Gerät. Ein Werkzeug aus Porzellan in ungewöhnlicher und ansprechender Form. Überzeugendes Produkt.

—

Judges panel

An innovative way of sharpening a knife—one that differs from the traditional knife sharpeners. A device that is always at hand on the worktop. A tool made of porcelain, in an unusual and appealing form. Impressive product.

—

Schaerfhut

Messerschärfer
Knife sharpener

Hersteller / Manufacturer
Mayer und Thiele GbR
D-42115 Wuppertal
—
Design / Designer
Vitamin D Design Services
D-42115 Wuppertal
—
Vertrieb / Distributor
Mayer und Thiele GbR
D-42115 Wuppertal
—

Für die gute Zubereitung von Speisen braucht man scharfe Messer. Wichtig ist dabei, sie vor und während des Kochens zu wetzen. Dafür ist der »Schaerfhut« ideal. Er steht sicher und rutschfest immer in greifbarer Nähe und kann mit einer Hand bedient werden.

Der »Schaerfhut« verbindet traditionelle Arbeits- und Funktionsweisen mit modernem Design. Die Form verschmilzt mit der Funktion und visualisiert den sparsamen Material- und Energieverbrauch. Der Winkel der kegelförmigen Schärffläche ist von allen Seiten benutzbar und so gewählt, dass die Messerklinge einfach nur senkrecht mit leichtem seitlichen Druck daran vorbeigeführt wird. Man erhält automatisch den optimalen Schärfwinkel von 10 bis 15 Grad.

Der »Schaerfhut« besteht aus einem durchgehenden Porzellankörper, vier kleinen Standfüßen aus Polyurethan und einem Klingenschutzring aus Polyamid, der verhindert, dass die Klinge großer Messer beschädigt wird. Porzellan ist wegen seiner Härte besonders geeignet zum Wetzen. Der Grundschliff wertvoller Messer bleibt erhalten.

Anyone who wants to prepare food properly needs a sharp knife. What is important here is to whet the knife before and during cooking. The »Schaerfhut« is ideal for this task. Non-slip, and always close by when it is needed, it nonetheless does not need both hands to use.

The »Schaerfhut« combines traditional methods and functions with modern design. It fuses form and function, and is a perfect illustration of how materials and energy can be used economically. The conical sharpening surface can be used from all sides, and has been set at an angle that allows the user to pass the knife along it vertically, applying slight pressure from the side. This automatically gives the optimum sharpening angle of between 10 and 15 degrees.

The »Schaerfhut« is made up of a monolithic porcelain body, four small polyurethane feet and a polyamide blade protector ring, which keeps the blades of large knives from damage. Porcelain's hardness makes it ideal for whetting. It helps maintain the original ground surface of valuable knives.

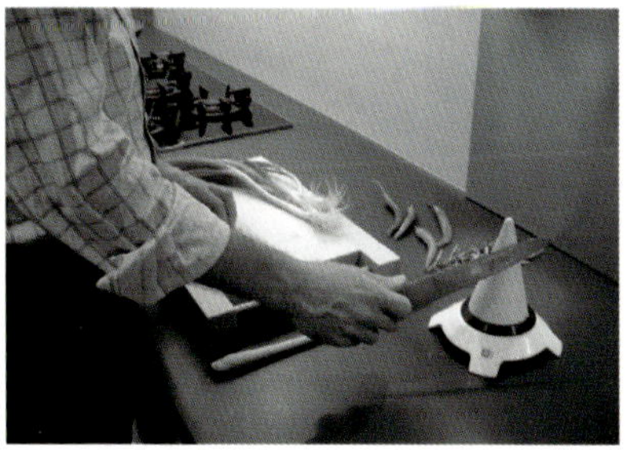

Quick-Chef 2
Küchengerät / Kitchen appliance

Küche, Haushalt / Silber
Kitchen, household / Silver

Quick-Chef 2

Küchengerät
Kitchen appliance

Hersteller / Manufacturer
Tupperware Belgium N.V.
B-9300 Aalst
—
Design / Designer
Jan-Hendrik de Groote und Vincent Jalet
B-9300 Aalst
—
Vertrieb / Distributor
Tupperware Deutschland GmbH
D-60488 Frankfurt
—

Der »Quick-Chef 2« ist die ideale Alternative zur elektrischen Küchenmaschine. Da das Produkt rein mechanisch arbeitet, wird bei der Benutzung nicht nur Strom gespart, sondern man kann es vollkommen unabhängig von der Steckdose, wie zum Beispiel beim Camping oder im Garten, einsetzen.

Der Schneideinsatz und der Rühreinsatz machen dieses besondere Küchengerät sehr vielseitig. So lässt sich damit nicht nur Obst oder Gemüse zerkleinern, sondern auch Sahne schlagen, Salatdressings, Dips oder Mayonnaise mixen. Der spezielle Kurbeldeckel übersetzt die Drehbewegung der Hand um ein Vielfaches auf die Einsätze, so dass das Zerkleinern und Rühren besonders einfach geht.

Der rutschfeste Ring am Behälterboden sorgt für einen sicheren Stand. Zum »Quick-Chef 2« gehören außerdem ein Sicherheitsverschluss und ein Trichter mit Messskala, mit dem man feindosiert Öl für Mayonnaise, Remoulade oder Dressings hinzugeben kann, ohne den Deckel immer wieder abnehmen zu müssen. Die Schüssel lässt sich mit einem zusätzlichen Deckel auch als Vorratsbehälter nutzen.

»Quick-Chef 2« is the ideal alternative to the electric food processor. Since the product works purely mechanically, it is not only an appliance that saves electricity, but also one that can be used completely independently of the mains socket—when camping, for example, or in the garden.

The blade and mixer sets provided with this unusual appliance make it very versatile. For example, it can not only chop fruit or vegetables, but also whip cream or mix salad dressings, dips or mayonnaise. The specially designed crank built into the lid multiplies one turn by hand into several turns of the blades or mixers, making chopping or mixing especially simple.

The non-slip ring fixed to the base of the bowl helps it to stand firm. »Quick-Chef 2« comes complete with a safety lock and a funnel with a measuring scale, which allows the user to add the exact amount of oil to mayonnaise, remoulade or dressings without having to constantly remove the lid. A second lid allows the bowl to be used for storing food.

Jury
Für das Thema »Focus Energy« ist es eine wunderbare Idee, dass man Geräte ohne elektrischen Strom bedienen kann, dass man seine eigene, statt der Energie aus der Steckdose nutzt. Einfaches Küchenwerkzeug, das durch seine praktischen Möglichkeiten überzeugt. Die Schneid- und Mixmechanik lässt sich abnehmen und durch einen luftdichten Deckel ersetzen, so dass der Behälter gleichzeitig als Vorratsbehälter genutzt werden kann.
—

Judges panel
For the »Focus Energy« topic, it is a wonderful idea to have appliances that can be used without electricity, and to use your own energy instead of energy from the plug. Simple kitchen tool with impressive practical applications. The cutting and mixing mechanism can be removed and replaced with an airtight lid, which means that the bowl can also be used for storing food.
—

Jury

Das gleiche Prinzip wie bei der größeren Maschine. Sehr praktischer Küchenhelfer, der statt elektrisch von Hand betrieben wird. Er ist einfach zu zerlegen, leicht zu spülen und zu reinigen, also ein überzeugendes Produkt.

—

Judges panel

The same principle as we saw in the larger appliance. Very practical kitchen aid, which is driven by hand instead of by electricity. It is easy to take apart, simple to wash and clean—in other words, a product that makes sense.

—

Happy-Chef

Küchenhelfer zum Schneiden
Chopping aid for the kitchen

Hersteller / Manufacturer
Tupperware Belgium N.V.
B-9300 Aalst
–
Design / Designer
Jan-Hendrik de Groote und Vincent Jalet
B-9300 Aalst
–
Vertrieb / Distributor
Tupperware Deutschland GmbH
D-60488 Frankfurt
–

Der »Happy-Chef« zerkleinert nicht nur Zwiebeln und Knoblauch schnell und ohne Tränen, sondern auch kleinere Mengen von Gemüse, Obst oder Kochschinken. Er arbeitet ohne Strom rein mechanisch. Damit kann er auch an Orten unabhängig von Steckdosen eingesetzt werden.

Der handliche Küchenhelfer besteht aus einem transparenten 300-ml-Behälter, dessen Boden mit einem gummiartigen Material ausgestattet ist und somit nicht rutschen kann.

Er ist sehr bequem in der Handhabung: Deckel und Behälter werden einfach entgegengesetzt vor- und zurückgedreht, so dass sich der Schneideinsatz bewegt. Der Drehdeckel liegt dank seiner Griffmulden gut in der Hand. Seine Zahnradmechanik überträgt die Drehbewegung um ein Mehrfaches auf die Klinge. An der Deckelunterseite befindet sich ein gummiartiger Rand, damit Behälter und Deckel gut miteinander verschlossen sind. Der Schneideinsatz mit zwei versetzt positionierten Edelstahlklingen, die beidseitig geschliffen und sehr scharf sind, zerkleinert einfach und schnell Obst oder Gemüse in beiden Drehrichtungen.

»Happy Chef« finely chops not only onions and garlic fast and without tears, but also small quantities of vegetables, fruit, or cold meat. It is completely mechanical, and works without electricity. This means it can be used anywhere, independently of any electricity supply.

This handy kitchen aid is made up of a transparent, 300 millilitre-capacity container with a rubber-like base, which means that it cannot slip.

It is very convenient to use. The lid and the bowl are simply moved to and fro against each other, and this causes the blades to move. The depressed grips in the turning lid mean that it fits well into the hand. The gear mechanism fitted into the lid multiplies its turning motion, so that the blades turn several times faster. The lower surface of the lid has a rubber-like edge, which ensures that it fits tightly onto the bowl. The blade unit comprises two staggered stainless steel blades. These are ground on both sides and are extremely sharp. As the lid is turned backward and forward, fruit and vegetables are chopped quickly and simply.

Fissler Intensa

Kochtopf
Saucepan

Hersteller / Manufacturer
Fissler GmbH
D-55743 Idar-Oberstein
—
Design / Designer
Milani design + consulting
CH-8703 Erlenbach
—
Vertrieb / Distributor
Fissler GmbH
D-55743 Idar-Oberstein
—

Das Kochsystem besticht durch sein gelungenes Design und durch viele praktische neue Funktionen. So zeigt die innovative Temperaturanzeige ThermoStar an, wann man den Herd zurückschalten kann: Die Farbe ändert sich von blau für kalt bis rot für heiß. Durch seine zwei unterschiedlichen Temperaturringe ist der ThermoStar optimal auf alle Herdarten abgestimmt.

Ebenfalls neu ist, dass der Deckel seinen eigenen Platz in der integrierten Deckelablage hat. Oder die im Topf- und Deckelrand integrierte Abgießfunktion, dank der man problemlos bei aufliegendem Deckel etwa Nudeln abgießen kann. Diese Funktion hat noch weitere Vorteile. Je nach Deckelstellung lässt sich der Topf zum vitaminschonenden Garen oder zum Abdampfen beziehungsweise Anbraten nutzen.

Der CookStar-Allherdboden, der für alle Herdarten geeignet ist, sorgt für perfekte Bodenstabilität und garantiert beste Wärmeaufnahme. Durch die leicht konische Form lassen sich alle Töpfe – auch gleichen Durchmessers – Platz sparend im Küchenschrank stapeln.

What is striking about this cooking system is its good design and a number of practical new functions. The innovative ThermoStar temperature display, for example, shows when cooking heat can be reduced. Its colour changes from blue for cold to red for hot. Its two different temperature rings mean that ThermoStar is perfectly compatible with any type of cooker.

What is also new is that the lid has its own place in the integrated lid holder. Or there is the pouring function integrated into the edge of the saucepan and the lid, which allows water to be drained away easily (when cooking pasta for example) even when the lid is on. This function offers even more advantages. Depending on the position of the lid, the saucepan can be used for gentle cooking, steaming, or frying.

The CookStar universal base, which is suitable for every type of cooker, provides perfect stability and optimum heat absorption. The saucepans' slightly conical form means that they can all be stacked together, even if their diameter is the same, thus saving space in the kitchen cupboard.

Jury
Ein auf den ersten Blick konventioneller Kochtopf, der allerdings durch seinen speziellen Griff besticht. Der Griff macht es möglich, dass der Deckel seitlich aufgestellt werden kann und so nicht die Arbeitsfläche vertropft. Er hat zudem eine Temperaturanzeige, die über Farbveränderung anzeigt, welche Temperatur im Topf herrscht. Ein ästhetisch überzeugender Topf.
—
Judges panel
At first glance, a conventional saucepan, though its special handle immediately stands out. This handle allows the lid to be held on the side of the pan, and so keeps drips away from work surfaces. It also has a temperature display which, by changing colour, shows the temperature inside the pot. An aesthetically pleasing saucepan.
—

AUERHAHN

Velvet

Besteck
Cutlery

Hersteller / Manufacturer
Auerhahn Bestecke GmbH
D-72213 Altensteig
—
Design / Designer
Cairn Young
GB London SE 17 3LJ
—
Vertrieb / Distributor
Auerhahn Bestecke GmbH
D-72213 Altensteig
—

Das Besteck besticht durch sein elegantes skulpturales Design, seine besonders schmeichelnde Haptik und seine einzigartige Ergonomie.

So bietet »Velvet« ein unvergleichliches sinnliches Handschmeichler-Erlebnis. Im Einklang mit seinen Proportionen schmiegt es sich über wohlgeformte Rundungen des samtigen, mattierten Edelstahls in die Hände. Es besitzt keine Ecken und Kanten, sondern nur sanfte Übergänge.

Sehr reizvoll ist dabei das Wechselspiel von polierten Innenlaffen und spiegelpolierten Klingen mit mattierten Außenlaffen und mattierten Stielen. Charakterisiert wird das Besteck auch durch seine signifikant ausgeformte Seitenlinie und die massive, wertvolle, elegante Anmutung dank hoher Materialstärken.

The outstanding features of this cutlery series are its elegant sculptural design, its especially pleasant tactile qualities and its unique ergonomic properties.

»Velvet« is extraordinarily sensual to the touch. In harmony with its proportions, the carefully crafted curves of its velvety, matt stainless steel fit perfectly in the hand. There are no sharp edges, but only gentle transitions.

What is especially attractive here is the interplay between the polished surfaces on the inside of the spoon bowls and the knife blades, and the matt surfaces on the outside of the spoon bowls and the knife handle. Other characteristic features of this cutlery include the prominent moulded line down the side and the solid, high-quality, elegant impression caused by the thickness of the materials used.

Jury
Ein Besteckset mit auf den ersten Blick konservativer Formgebung, das aber über seine ergonomischen Qualitäten und seine eigene Zeichensprache sehr überzeugt hat. Sehr schön die polierten und die matten Bereiche.
—
Judges panel
A set of cutlery whose design appears conservative at first glance. However, its ergonomic properties and inimitable semiotics soon show that it is precisely the opposite. The idea with the polished and matt areas is very good.
—

Küche, Haushalt / Silber
Kitchen, household / Silver

Für Zwei
Mehrzweckmesser / Multi-purpose knife

Für Zwei

Mehrzweckmesser
Multi-purpose knife

Hersteller / Manufacturer
Windmühlenmesser
Robert Herder GmbH & Co. KG
D-42697 Solingen
—
Design / Designer
Werksdesign / In-house design
Giselheid Herder-Scholz
—
Vertrieb / Distributor
Windmühlenmesser
Robert Herder GmbH & Co. KG
D-42697 Solingen
—

Trotz seiner mit 3,5 Zoll relativ kurzen Klinge ist »Für Zwei« ein vollwertiges Parier- und Kochmesser auf dem Brett. Das Geheimnis liegt in seinem langen Griff. Wegen der durchgehenden Klinge hat es einen etwas schwerer ausgewogenen Griff als andere Küchenmesser.

Das Mehrzweckmesser arbeitet wie ein traditionelles Küchenmesser – nur viel schärfer. Feines Gemüse oder Obst schält es hauchdünn und das Würfeln einer Zwiebel oder Möhre geht so präzise, dass man den Widerstand kaum spürt. Durch seinen enorm spitzen Schneidwinkel erzielt es auch beim Schälen eine saubere und ästhetische Schnittfläche.

Die Griffe sind handgefertigt aus hochwertigem fein gemasertem Oliven-, Pflaumen- oder Birnenholz, von Meisterhand feingeschliffen, poliert und geölt. Die Klinge besteht aus rostfreiem Stahl.

Despite its relatively short blade (just 3.5 inches), »Für Zwei« is a fully functional paring and cooking knife. Its secret is in its long handle. As the tang runs the whole length of the handle, the handle is slightly heavier and better balanced than that of other kitchen knives.

The multi-purpose knife works like a traditional kitchen knife, but is much sharper. It can cut away the finest of peel from vegetables or fruit, and chops a carrot or an onion so precisely that hardly any resistance can be felt. Its extremely pointed cutting angle means that it creates a clean, aesthetic cut edge, even when peeling.

The handles are hand-made of high-quality, fine-grained olive, plum or pear wood, finely sanded, polished and oiled by master craftsmen. The blade is made of stainless steel.

Jury
Praktisches, angenehmes klassisches Gemüse- oder Mehrzweckmesser. Unterschiedliche europäische Hölzer. Insofern ein sehr ökologischer Ansatz.
—
Judges panel
Practical, pleasantly classical vegetable knife or multi-purpose knife. Uses various European types of wood. This is a very ecological approach.
—

Cool!

Tranchierbesteck
Carving set

Hersteller / Manufacturer
Eva Denmark A/S
DK-2760 Måløv
—
Design / Designer
Claus Jensen & Henrik Holbæk
Tools®
DK-2400 Kopenhagen NV
—
Vertrieb / Distributor
Eva Denmark Vertriebsleitung Deutschland
Thomas Schmidt
D-56112 Lahnstein
—

Das elegante, puristische und ganz reduziert gestaltete Tranchierbesteck ist eine Zierde auf jedem Esstisch. Die Form ist so kultiviert, cool und unglaublich einfach, dass es schade wäre, den Braten in der Küche statt direkt auf dem Esstisch oder beim Grill zu tranchieren.

Das Tranchierbesteck besteht aus einer Gabel mit zwei Zinken und einem ultrascharfen Messer. Es ist aus asiatischem Messerstahl geschmiedet und bis zur Perfektion geschärft. Messer und Gabel sind jeweils aus einem einzigen Stück gefertigt, der Griff läuft nahtlos in Klinge oder Zinken über.

The elegant, reduced and uncompromisingly minimalist design of this carving set is an eye-catcher on any dining table. The shape is so sophisticated, cool and unbelievably simple that it would be a shame to slice the roast in the kitchen instead of at the dining table or the barbeque.

The carving set comprises a two-pronged fork and a super-sharp carving knife. It is made of Asian cutlery steel and is sharpened to perfection. The knife and fork are each crafted from one piece with the handle running seamlessly through to the blade or prongs.

Jury
Gute Kombination aus innovativer Form und traditionellem skandinavischem Purismus. Sehr ausgewogem im Gewicht, sehr gute Ergonomie. Exzellenter Einsatz des Materials, aber auch ein gutes expressives Werkzeug zum Tranchieren, wenn man Gäste hat. Es ist sowohl ein schönes als auch ein nützliches Produkt.
—
Judges panel
A good combination of an innovative shape and traditional Scandinavian minimalism. Very well balanced, extremely ergonomic. Wonderful use of material but also the perfect statement when carving before guests. A beautiful and useful product in one.
—

Summersalad

Salatbesteck
Salad set

Hersteller / Manufacturer
Eva Denmark A/S
DK-2760 Måløv
—
Design / Designer
Claus Jensen & Henrik Holbæk
Tools®
DK-2400 Kopenhagen NV
—
Vertrieb / Distributor
Eva Denmark Vertriebsleitung Deutschland
Thomas Schmidt
D-56112 Lahnstein
—

Das ungewöhnliche Salatbesteck sticht durch seine überdimensionierten Löffel und Gabeln ins Auge. Sie werden auf jedem Tisch zu einem Blickfang. Das Material, rostfreier Edelstahl, und die außergewöhnliche, auch humorvolle Gestaltung geben dem Salatbesteck seinen besonderen Charakter.

Das Salatbesteck ist gut in der Hand zu halten. Die tiefe Form des Löffels und der Gabel macht es leicht und bequem, Salat von einem Teller auf den anderen zu befördern.

The unusually large fork and spoon in this salad set put it apart from the crowd. A real centrepiece on any table. The stainless steel material and the unusual, witty design give this salad set a very special character.

The salad set lies well in the hand. The depth of the spoon and the fork makes it easy to pass cumbersome salad from one plate to another.

Jury
Sehr spielerisches Besteck, das uns von der Art, wie wir über Industriedesign denken, befreit. Hat sehr viel Humor. Erinnert an das Spiel im Sandkasten oder am Strand. Was davon im Erwachsenenalter bleibt, ist das Spiel mit dem Salat in der Salatschüssel. Dies ist ein sehr interessantes Objekt, weil es eine Geschichte und einen Erinnerungswert hat. Ein neuer Aspekt im Design.
—
Judges panel
Very playful salad set which frees us from the way we think about industrial design. It is very witty. It reminds one of childhood days in the sandpit or at the beach. For adults this memory is relived when playing with salad in the salad bowl. This is an extremely interesting piece of design because it has a history and a sentimental value. A new aspect in design.
—

Küche, Haushalt / Silber
Kitchen, household / Silver

Mixxo
Stabmixer / Hand blender

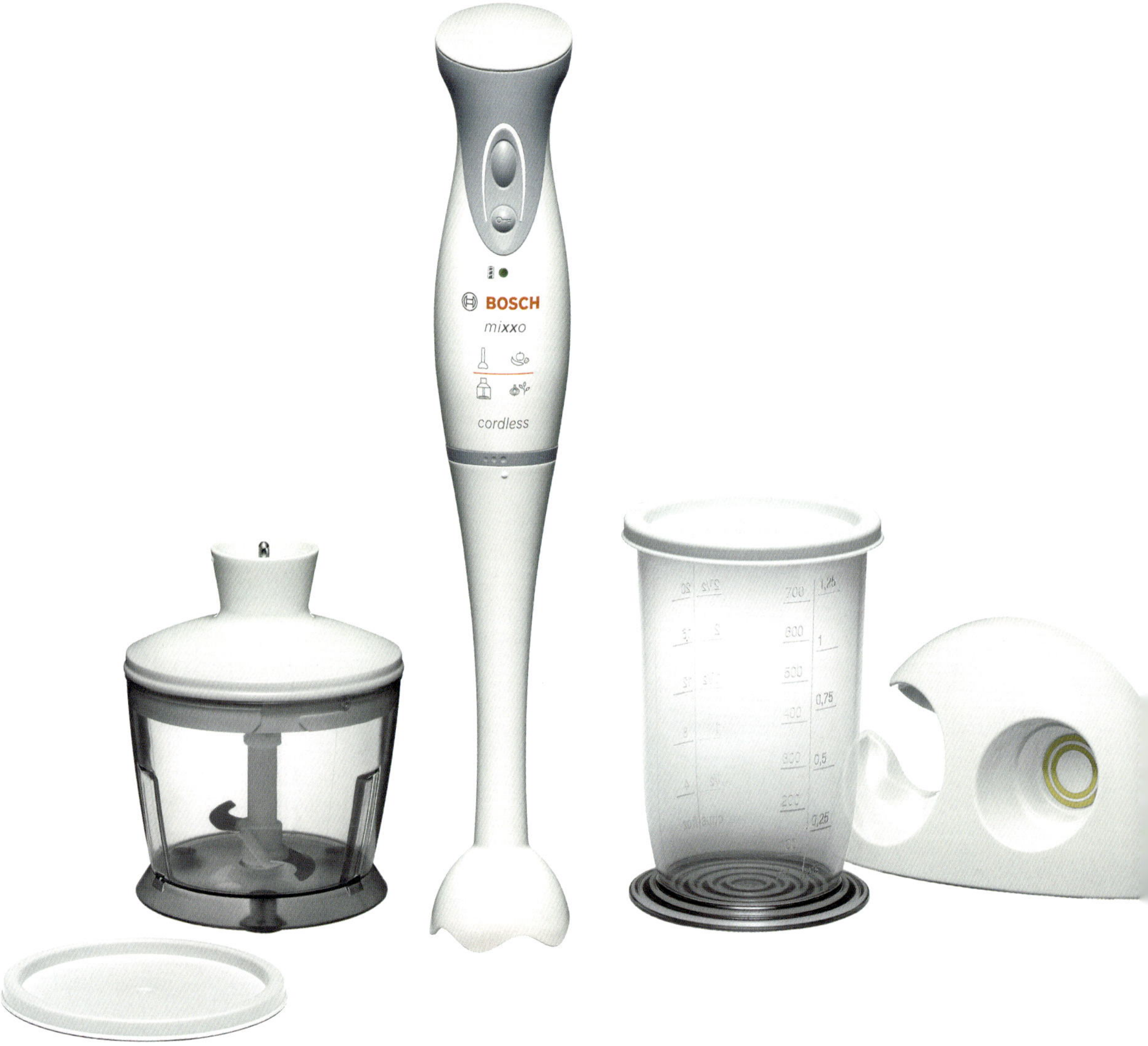

Mixxo

Stabmixer
Hand blender

Hersteller / Manufacturer
Robert Bosch Hausgeräte GmbH
D-81739 München
—
Design / Designer
Daniels & Koitzsch
D-64285 Darmstadt
—
Vertrieb / Distributor
Robert Bosch Hausgeräte GmbH
D-81739 München
—

Mit dem weltweit ersten kabellosen Stabmixer mit Lithium-Ionen-Akkus kann man überall zerkleinern, pürieren und mixen. Das speziell entwickelte Edelstahlmesser mit zwei zusätzlichen, vertikalen Vorschneidern hat selbst bei harten Zutaten, wie Nüssen, keine Probleme.

Der »Mixxo« ist klein, leicht und handlich, sein Mixfuß ist jedoch extra lang, damit man auch in tiefen Töpfen pürieren kann. Gereinigt wird der Fuß einfach in der Spülmaschine.

»Mixxo« besticht durch sein schlankes, elegantes Design. Sein ergonomisch geformter Griff mit den Soft-Touch-Griffeinlagen und seine ausgewogene Gewichtsverteilung erlauben eine komfortable Handhabung – und das gleichermaßen für Links- wie Rechtshänder. Für zusätzliche Sicherheit sorgt die Zwei-Tasten-Sicherheitsschaltung.

Die Ladestation mit LED-Ladestandsanzeige kann sowohl auf dem Tisch aufgestellt als auch an der Wand montiert werden. Besonders praktisch ist das Zubehör, wie der Universalzerkleinerer und der Mixbecher, in dem selbst Eiswürfel zerstoßen werden können.

The world's first cordless hand blender, powered by lithium-ion rechargeable batteries. Now food can be chopped, blended and mixed anywhere. The specially developed stainless steel knife, with two additional, vertical pre-cutters, easily processes even the hardest ingredients, such as nuts.

Although »Mixxo« is small, light and handy, its blending shaft is extra long, and so can reach the bottom of even the deepest saucepans. It can be detached from the handle and washed in the dishwasher.

»Mixxo« is exceptionally slim and elegant in design. Its ergonomically shaped handle with soft touch grip areas and good balance means that it is very comfortable to use—and it is equally suitable for left and right-handers. The dual-button safety switch provides extra safety.

The recharging base with its LED charge status display can be used on the table top or mounted on the wall. It comes with especially practical accessories, such as a universal chopper and a mixing beaker in which even ice cubes can be pulverized.

Jury
Praktisches Küchenutensil, spülmaschinengeeignet, mit dem kabellosen Betrieb sehr flexibel, sehr breit einsetzbar, modularer Aufbau. Gute Lösung. Überzeugendes Design.
—
Judges panel
Practical kitchen utensil, dishwasher-safe, very flexible thanks to cordless operation, very versatile application, modular design. Good solution. Convincing design.
—

Unit 2

Küchenarbeitsplatz
Kitchen workplace

Hersteller / Manufacturer
Eisfink Max Maier GmbH & Co. KG
D-71636 Ludwigsburg
—
Design / Designer
Pearl Creative
Tim Storti, Christian Rummel
und Max Maier
D-71636 Ludwigsburg
—

»Unit 2« vereint die grundlegenden Funktionen einer Küche – Wasser-, Arbeits- und Kochstelle – in einem einzigen hochfunktionalen Modul. Entstanden ist eine an gewerbliche Küchen angelehnte, ergonomisch optimal und zeitlos gestaltete Arbeitsplattform, die frei im Raum aufgestellt werden kann und Profiqualität in die Privatküche bringt.

So wurden das Konzept der Spüle und die Absaugtechnik aus dem gewerblichen Bereich übernommen. Die im Hauptbecken integrierte zweite Stufe ermöglicht den perfekten Umgang mit Gastronormbehältern, die als Industriestandard verfügbar sind. Die große GN-Schale kann – um 90° gedreht – auch auf der unteren Ebene eingehängt werden. Ebenso lassen sich das GN-genormte Schneidbrett und kleinere GN-Behälter einhängen. Durch die vielen verschiedenen Beckengrößen lässt sich der Wasserverbrauch deutlich minimieren.

Der patentierte Air Clean 3D-Abzug saugt die fetthaltige, verschmutzte Luft unmittelbar an der Kochstelle ab und führt die warme Luft energiesparend und geruchsfrei wieder in den Raum zurück.

»Unit 2« brings together the basic functions of a kitchen—watering place, workplace and cooking place—to make one single, highly functional module. The result is a work platform that borrows from the world of commercial kitchens, and that is ergonomically optimized and timeless in design. It can stand on its own in a room, and lend an air of professionalism to a private kitchen.

The idea behind the washbasin and the suction technique used is borrowed from the commercial sphere. The second level integrated in the main basin allows Gastro Norm (GN) containers to be integrated perfectly. These containers are available as the industry standard. If it is turned 90 degrees, the large GN bowl can also be suspended on the lower level. The GN-standard cutting board and smaller GN containers can also be suspended here. The many different basin sizes mean that water consumption can be significantly reduced.

The patented Air Clean 3D extractor fan draws away airborne fat and fumes directly at the cooker, returning the warm air to the room free of odour.

Jury
Ein modulares System mit einer sehr effizienten Produktionsstrategie. Sehr gute Lösung: Durch die verschieden tiefen Becken lässt sich der Wasserverbrauch reduzieren. Die Dunstabzugshaube gewinnt die Energie in Form von Wärme zurück, das heißt, es wird keine beheizte Luft ins Freie abgesaugt. Ein sehr energieeffizientes Konzept.
—
Judges panel
A modular system with a very efficient product strategy. Very good solution: because the washbasins are different depths, water consumption can be reduced. The extractor hood recovers energy in the form of heat, and so no heated air is expelled into the outside atmosphere. A very energy-efficient concept.
—

Waterstation cubic 980
Küchenspüle / Kitchen sink

Waterstation cubic 980

Küchenspüle
Kitchen sink

Hersteller / Manufacturer
Eisfink Max Maier GmbH & Co. KG
D-71636 Ludwigsburg
—
Design / Designer
Pearl Creative
Tim Storti, Christian Rummel
und Max Maier
D-71636 Ludwigsburg
—

Die »Waterstation cubic 980« ist die erste Spüle, die aus gezogenem Edelstahl besteht und dabei dank der sehr kleinen Radien den gestalterischen Ansprüchen einer modernen Küche entspricht. Die Spüle, die professionelle Qualität in den Privathaushalt bringt, hat so eine moderne, kubische Anmutung bekommen, lässt sich aber dennoch in den Ecken gut reinigen.

In Profiküchen gibt es weltweit ein Produkt, das alle Geräte zum System verbindet: Der Gastronormbehälter mit dem rechteckigen Grundmaß von 530 x 325 Millimeter und zahlreichen Unterabmessungen.

Diese GN-Behälter lassen sich auch bei der »Waterstation cubic 980« auf zwei Arbeitsebenen einsetzen: entweder auf der Höhe der Arbeitsfläche oder tiefer, auf der zweiten eingezogenen Stufe im Becken. Die Lösung ermöglicht effizientes, einfaches Zubereiten von Speisen, da es nun möglich ist, mit einer Vielzahl von Zubehörteilen auf verschiedenen Ebenen zu arbeiten. Das spart Platz auf der Arbeitsfläche und ist ergonomisch sinnvoll für unterschiedlich große Nutzer.

Jury
Sehr guter modularer Ansatz: in der edelstahlgezogenen Spüle kann man verschiedene Behälter auf zwei Höhen einsetzen. Sehr praktisch. So kann man oben Gemüse schneiden und die fertig geschnittenen Stücke gleich in einen Behälter darunter schieben und dort waschen. Auch ohne größeren Aufwand zu produzieren, im Sinne des Energieansatzes sehr überzeugend.
—
Judges panel
Excellent use of a modular approach: a range of containers can be used at two different levels in this deep drawn stainless steel sink. Very practical. For example, vegetables can be diced at the top level and pushed straight into the container below where they can be washed. Can also be produced without great expense, and very convincing in terms of the use of energy.
—

The »Waterstation cubic 980« is the first kitchen sink made of drawn stainless steel, a technique that allows use of tight radii. The result lives up to the exacting design standards expected of a modern kitchen. This sink unit, which brings professional quality into the private kitchen, has a modern cubist feel to it but can still be cleaned well in the corners.

In professional kitchens there is one product worldwide which combines all utensils in just one system: the standard Gastro Norm container, measuring 530 x 325 mm and numerous smaller variations.

This GN container can also be used in the »Waterstation cubic 980« on two levels: either at benchtop level or further down on a second, recessed level in the sink itself. The solution allows efficient and simple preparation of food, as it is now possible to work with numerous accessories on different levels. This saves space on the bench and makes ergonomic sense no matter how tall the user may be.

Varithekservo+

Servierwagen
Serving trolley

Hersteller / Manufacturer
Rieber GmbH & Co. KG
D-72770 Reutlingen
—
Design / Designer
Pearl Creative
Tim Storti, Christian Rummel
und Max Maier
D-71636 Ludwigsburg
—
Vertrieb / Distributor
Rieber GmbH & Co. KG
D-72770 Reutlingen
—

Die Systemfamilie »Varithekservo+« hebt sich völlig von klassischen Servierwagen ab. Die bisherigen voluminösen Schweißkonstruktionen wurden durch ein zerlegbares, geschraubtes System ersetzt. Damit ist es gelungen, die Stabilität und so die Langlebigkeit gegenüber geschweißten Ausführungen um den Faktor 10 zu erhöhen.

Die neugewonnene Zerlegbarkeit bietet noch weitere Vorteile. Der Servierwagen lässt sich kinderleicht in weniger als 10 Minuten montieren. Das Volumen des bisherigen Packmaßes hat sich auf ein Viertel reduziert. Damit lassen sich Transport- und Lagerkosten, vor allem aber der Energieverbrauch beim Transport erheblich senken.

»Varithekservo+« überzeugt auch durch sein zeitgemäßes Erscheinungsbild. Weitere funktionale Komponenten aus der Varithek-Familie machen aus dem Standard-Servierwagen neue eigenständige Produkte: Warmhaltewagen, Kochwagen, Kühlwagen und Grillwagen. »Varithekservo+« wird damit zur mobilen Küche, wo Gäste zuschauen können, wie Speisen zubereitet werden.

The »Varithekservo+« product family is in a class of its own compared with traditional serving trolleys. The standard cumbersome welded constructions have here been replaced by a knock-down, bolted system. This has made it possible to increase the life and stability of the trolleys by a factor of 10 over those of welded models.

The knock-down feature also offers additional advantages. It can easily be assembled in less than 10 minutes. The packing volume has been reduced to a quarter of what it was. This saves not only transport and storage costs but—more importantly—energy consumption during transportation.

The modern design of »Varithekservo+« is compelling. Other functional components taken from the Varithek family turn the standard serving trolley into completely new standalone products: warming tray, portable kitchen, mobile refrigerator, and barbeque. The »Varithekservo+« is thus a mobile kitchen where the guests can watch the dinner being prepared.

Jury
In der heutigen Bürokultur mit langen Meetings wird zunehmend Essen in den Besprechungsraum gebracht. Gute Lösung, um heißes Essen zu transportieren und warmzuhalten. Überzeugend das um 75 % reduzierte Verpackungsvolumen, das den Transport effizienter macht und Energie einspart. Auch die Materialien, nämlich Edelstahl und einfach zu recycelnde Kunststoffteile, überzeugen.
—

Judges panel
In today's office culture of long meetings, it is becoming increasingly common for food to be brought into the meeting room. A good solution to transport hot food and keep it warm. The 75% reduction in packing volume, which makes transport more efficient and saves energy, is a compelling argument. Convincing choice of materials: stainless steel and easy-to-recycle plastics.
—

LeMans
Eckschrank-Schwenkauszug / Corner unit swivel shelves

LeMans

Eckschrank-Schwenkauszug
Corner unit swivel shelves

Hersteller / Manufacturer
Kesseböhmer GmbH
D-49152 Bad Essen
—
Design / Designer
Werksdesign / In-house design
—
Vertrieb / Distributor
Kesseböhmer GmbH
D-49152 Bad Essen
—

Der Eckschrank-Schwenkauszug »LeMans« bietet perfekten Zugriff bis in den letzten Winkel des Schrankinhalts. Die beiden Böden – vier bei Highboards – lassen sich einzeln mit einer leichten dynamischen Bewegung vollständig aus dem Schrankkorpus herausschwenken, so dass die gesamte Stellfläche zugänglich ist.

Die dafür entwickelten abgerundeten, asymmetrischen Böden erinnern an die Kurven der Rennstrecke von Le Mans, der der Schwenkauszug seinen Namen verdankt. Diese dynamische Form basiert auf der bestmöglichen Ausnutzung des Stauraums.

Die Kombination von Drehen und Schwenken ermöglicht einen eleganten und harmonischen Bewegungsablauf und den vollen und bequemen Zugriff außerhalb des Schrankes. Jeder Boden wird einzeln bewegt und kann bis zu 20 Kilo tragen. Die serienmäßige »Arenaplus Rutschstopp«-Beschichtung hält das Staugut sicher auf seinem Platz. Dazu gibt es wahlweise die Anschlagsdämpfung und einen Deckelhalter.

With its turning motion, the »Le Mans« corner unit makes it possible to reach even the very corners of the kitchen cupboard. With a gentle push, the two shelves — or four in the highboard unit — can be swivelled individually and completely out of the unit to grant easy access to the entire storage space.

The specially developed, rounded and asymmetric shelves are reminiscent of the curves of the Le Mans racetrack, as the name suggests. This dynamic shape is based on the best possible use of storage space.

The combination of a turning and swivelling motion allows complete and easy access to the contents of the unit in one elegant and harmonious movement, without the need to reach into the cupboard. Each shelf moves separately and can carry up to 20 kg. The standard »Arenaplus antislip« surface keeps the contents from slipping out of place. Optional extras include a buffer and a container lid.

Jury
Gute Lösung für die tote Ecke im Küchenschrank. Der zweifache Drehpunkt bietet Zugriff auf den kompletten Schrankinhalt. Optimale Ausnutzung des toten Raums.
—
Judges panel
Good solution for the wasted space in the corners of a kitchen unit. The double action turning motion offers access to the entire contents of the cupboard. Best possible utilization of the wasted space.
—

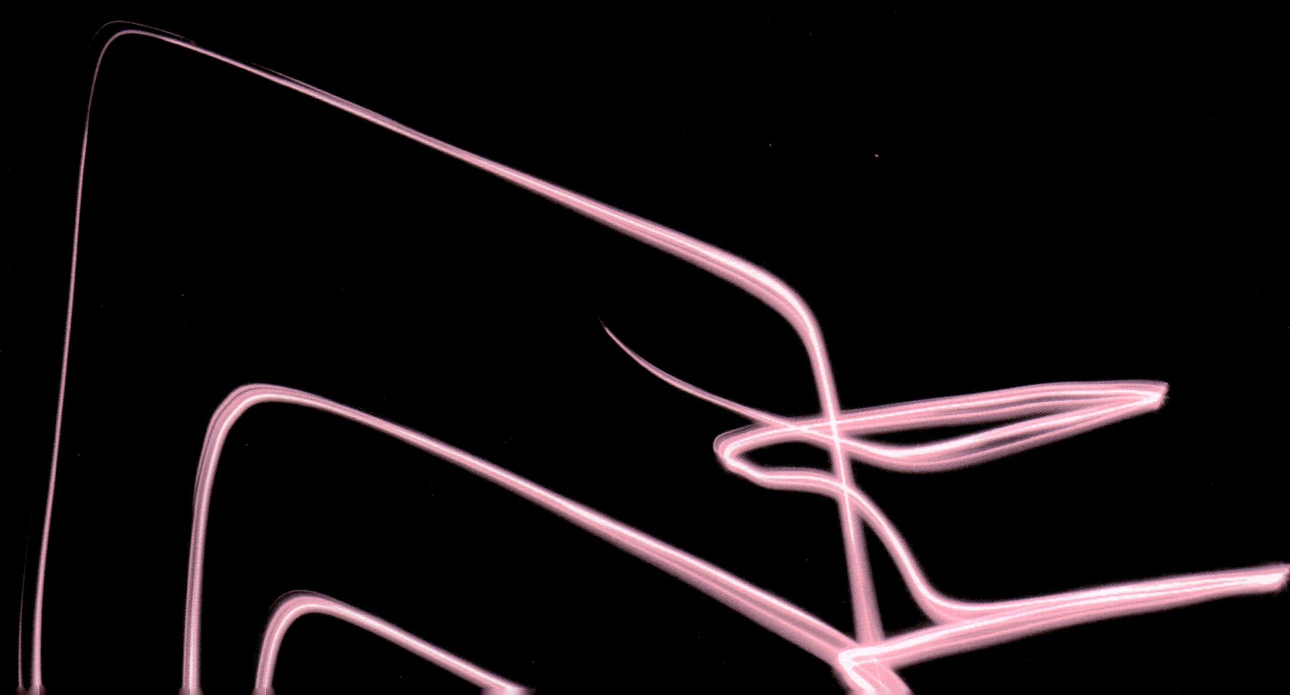

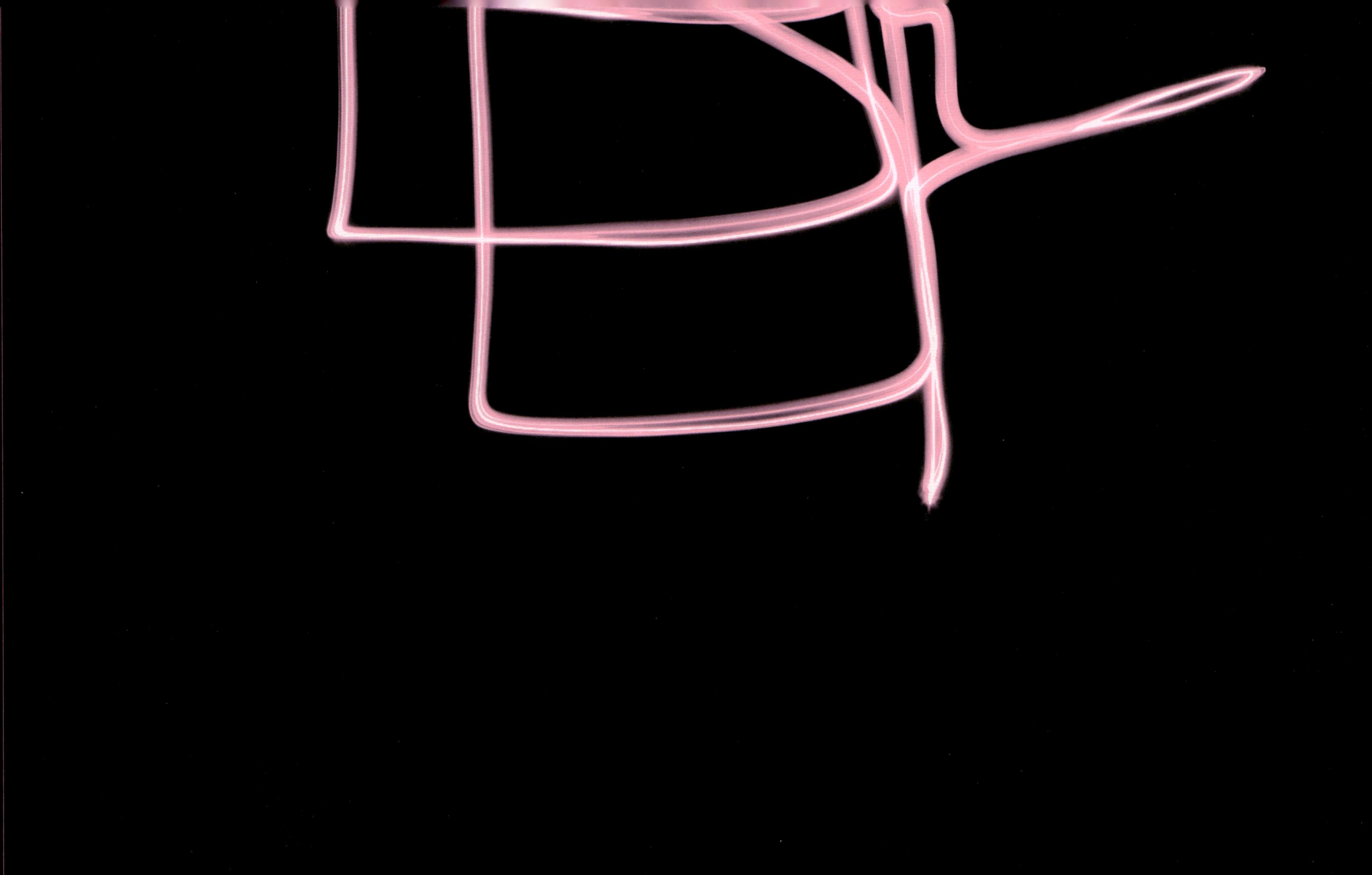

Aline
Stapelstuhl / Contract chair

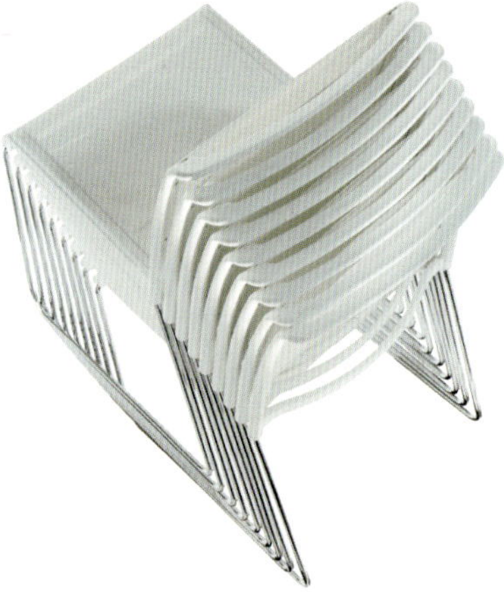

Aline
Stapelstuhl / Contract chair

Aline

Stapelstuhl
Contract chair

Hersteller / Manufacturer
Wilkhahn
Wilkening + Hahne GmbH + Co. KG
D-31848 Bad Münder
—
Design / Designer
Andreas Störiko
I-20123 Mailand
—
Vertrieb / Distributor
Wilkhahn
Wilkening + Hahne GmbH + Co. KG
D-31848 Bad Münder
—

Der leichte und trotzdem robuste Stapelstuhl »Aline« zeichnet sich durch minimierten Materialeinsatz und einfachste Handhabung aus. Der innovative Kufenstuhl ist ein ebenso filigraner wie komfortabler Mehrzweckstuhl, der sowohl für sich allein als auch in großen Reihen stehen und somit in ganz unterschiedlichen Kontexten genutzt werden kann. Das funktionale Design zeigt sich besonders eindrucksvoll in der prägnanten Ausbildung des Sitzrahmens nach dem »Kufen-Schiene-Prinzip«. Es ermöglicht ein formschlüssiges, äußerst kompaktes Stapeln.

Freitragende Sitz- und Rückenrahmen aus hochfestem, glasfaserverstärktem Polyamid, die mit einem leistungsfähigen, ummantelten Polyestergewebe bespannt sind, sind auf ein filigranes Kufengestell aus glanzverchromtem Rundstahl aufgebracht. Das Kufengestell bildet vorne die Auflage für den Sitzrahmen, hinten durchdringt es den Sitzrahmen und wird in den Rückenrahmen eingeführt. Gestell und Rahmen bilden dadurch ein stabiles Raumgitter mit geringen Materialdimensionen und niedrigem Gewicht. Die Bespannung der Rückenlehne passt sich jeder Körperform an und bietet dank Hinterlüftung ein angenehmes Sitzgefühl.

Minimal use of materials and exceptionally easy handling are the distinctive features of the »Aline« stackable chair. It is light but nevertheless robust. This novel skid chair, slender and comfortable, is a multipurpose chair that can either stand alone or in long rows, and thus be used in many different contexts. The functionality of the design is particularly apparent in the distinctive shape of the seat frame, on the »skid-rail« principle. It allows form-fitting, extremely compact stacking.

Floating seat and backrest frames of high-strength, glass fibre-reinforced polyamide, covered with a high-performance sheathed polyester fabric, are mounted on a slender skid-type support of high-gloss chrome-plated round-section steel. At the front the skid frame forms the support for the seat frame. At the rear it passes through the seat frame, extending into the backrest frame. The support and seat frame thus form a sturdy spatial framework with small dimensions of the materials used and low weight. The backrest covering adapts to any anatomical shape and is very comfortable, being ventilated from the rear.

Jury
Sehr gut gestaltete Struktur in Kunststoff zusammen mit Metall. Kaum Gewicht. Viele Verwendungsmöglichkeiten. Ungewöhnlich der sehr innovative Verbindungsmechanismus, den keiner der Juroren bisher gesehen hat. Er hat alle Funktionen, die ein stapelbarer und konferenzfähiger Stuhl braucht. Ein effizienter Ansatz, mit wenig Material viel zu erreichen. Sehr gut ausgeführt.
—
Judges panel
Very well designed plastic and metal construction, almost weightless. Wide range of applications. Very unusual, innovative connecting mechanism, never seen before by any of the judges. Has all of the functions necessary for a stackable contract chair. An efficient approach to achieving a lot with little material. Very well made.
—

Jury

Herausragender Bürodrehstuhl durch hervorragende Verarbeitung von Aluminium und Leder und auf Grund der einzigartigen Kinematik des Gestells. Außergewöhnlich die Verbindung zwischen dem Sitzteil und der Armlehne. Eindrucksvoll die Art, wie die Armlehne exakt an dem Radius des Sitzteils entlanggleitet.

—

Judges panel

Superb office swivel chair, the result of excellent workmanship in aluminium and leather and the unique kinematics of the frame. Unusual connection between seat and armrest. Impressive how the armrest glides so precisely along the radius of the seat element.

—

Silver

Bürodrehstuhl
Office swivel chair

Hersteller / Manufacturer
Interstuhl Büromöbel GmbH & Co. KG
D-72469 Meßstetten-Tieringen
–
Design / Designer
Hadi Teherani AG
Hadi Teherani, Hans Ulrich Bitsch, Ulrich Nether
D-20427 Hamburg
–
Vertrieb / Distributor
Interstuhl Büromöbel GmbH & Co. KG
D-72469 Meßstetten-Tieringen
–

»Silver« zeichnet sich durch eine elegante Schalenbauweise aus Aluminium aus. Ungewöhnlich sind die transparenten Gleiter sowie die Bedienelemente, die perfekt ins Design integriert und dennoch intuitiv und einfach mit einem einzigen Griff zu handhaben sind. Zudem wurden bei dem futuristisch anmutenden Bürodrehstuhl Karosserieelemente verwendet, die an Fahrzeugbau erinnern.

»Silver« charakterisiert sich durch neue Bewegungsabläufe und eine innovative, ablesbare Technik, etwa bei der Bewegung des Rückens. Er bietet ein Höchstmaß an Funktionalität und Ergonomie – und das ohne komplizierte und überflüssige Einstellprozeduren. Beispielsweise sind die Griffmulden in der Sitzschale zur Bedienung der Höhenverstellung und der Rückenlehnenarretierung perfekt integriert und dennoch jederzeit zu bedienen.

Man kann zwischen vier Lederfarben, sechs unterschiedlichen Stoffen und drei Aluminiumoberflächen wählen: pulverbeschichtet, gestrahlt oder gebürstet.

The distinctive feature of »Silver« is its elegant aluminium bucket-type construction. Unusual features are the transparent gliders and the controls, which are perfectly integrated into the design and nevertheless intuitively controllable—simply and with a single movement. This futuristic-looking office swivel chair has bodywork elements reminiscent of vehicle construction.

Characteristic features of »Silver« are new motion sequences and an innovative, readable technology, for example during movement of the back. It offers maximum functionality and ergonomy without any complex or superfluous setting procedures. For example, the finger-holes for adjusting the seat height and locking the backrest are perfectly integrated but can nevertheless be used whenever needed.

There is a choice of four colours of leather, six fabrics and three aluminium surface finishes: powder-coated, sandblasted or brushed.

Jury

Eine sehr gute Lösung, vereint die unterschiedlichsten Anforderungen an Bürostühle bezüglich Höhenverstellung und Seitenverstellung der Armlehnen. Auch die ergonomischen Anforderungen sind gut gelöst mit dem reduziert gestalteten Träger, an den die Rückenlehne angebracht wurde. Der Stuhl fügt sich neutral in jede Büroumgebung ein.

—

Judges panel

A very good solution; satisfies a very wide range of criteria for office chairs, with regard to height adjustment and lateral adjustment of the armrests. The ergonomic criteria are also well fulfilled, with the frugal design of the back section, to which the backrest is attached. The chair fits in neutrally with any office environment.

—

HeadLine

Bürodrehstuhl
Office swivel chair

Hersteller / Manufacturer
Vitra GmbH
D-79576 Weil am Rhein
—
Design / Designer
Mario Bellini mit Claudio Bellini
c/o Vitra GmbH
D-79576 Weil am Rhein
—
Vertrieb / Distributor
Vitra GmbH
D-79576 Weil am Rhein
—

»HeadLine« ermöglicht bei gleichzeitiger Unterstützung mehr Bewegungsfreiheit und verursacht so weniger Spannungen. Die innovative Lehne – zweigeteilter Rücken und hochflexible, verlängerte Rückenlehne – passt sich automatisch an Nutzer jeder Größe an und führt beim Zurücklehnen Schultern und Nacken. Die Wirbelsäule wird aufgerichtet und entlastet, der Kopf optimal gestützt.

Besondere Stabilität erhält die Rückenlehnenkonstruktion durch eine ausgeklügelte Innovation: Zwei Aluminiumholme sind mittels zweier Pendelgelenke mit dem flexiblen Rücken verbunden.

Neben der Rückenlehnenkonstruktion ist auch das Klimakonzept außergewöhnlich und innovativ, das für hohen Sitzkomfort sorgt. Beste Luftzirkulation garantiert der Bezug, der aus einem dreidimensionalen Abstandsgewirke besteht. Unter dem Sitzbezug nimmt der patentierte Klimasitz Feuchtigkeit auf und leitet sie rasch ab. Der Bürodrehstuhl wird aus 53 Prozent recyceltem Material hergestellt und ist zu 98 Prozent recycelbar.

»Headline« offers not only support, but also more freedom of movement, thus causing less tension. The innovative back support—with divided back and highly flexible, elongated backrest—adapts automatically to the size of the user, guiding the shoulders and nape when leaning back. The spine is straightened and relieved, the head ideally supported.

An ingenious innovation makes the backrest construction especially sturdy: two aluminium shafts are connected to the flexible back by means of two pendulum joints.

Besides the backrest construction, the microclimate concept is unusual and innovative, ensuring that the chair is pleasant to sit on, with a high level of comfort. The covering, which consists of a three-dimensional spacer fabric, guarantees optimum air circulation. Beneath the seat covering the patented climate seat absorbs and quickly dissipates moisture. Fifty-three per cent of the chair consists of recycled material; it is 98 per cent recyclable.

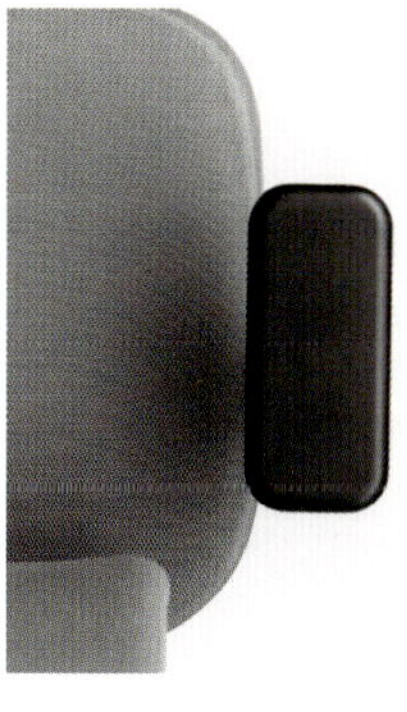
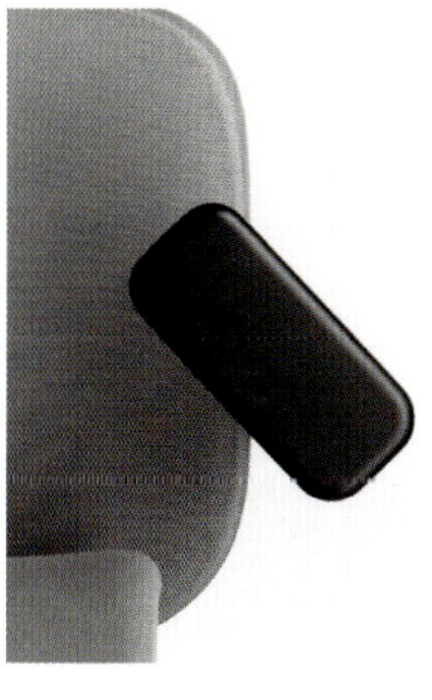

Drift

Sessel
Armchair

Hersteller / Manufacturer
Walter Knoll AG & Co. KG
D-71083 Herrenberg
—
Design / Designer
EOOS Design GmbH
A-1010 Wien
—
Vertrieb / Distributor
Walter Knoll AG & Co. KG
D-71083 Herrenberg
—

Der Sessel ist ein markanter Solitär. Mit einem Drehteller ausgestattet, wird »Drift« ganz leicht zum »Bewegungssessel«. Schwungvoll kann man sich seinem Gesprächspartner zuwenden, mit leichten Links- und Rechts- Bewegungen dynamisch sitzen oder einfach nur hin und wieder die Blickrichtung ändern.

Die organischen Polsterflächen gestatten außerdem ganz verschiedene Sitzpositionen: aufrecht und konzentriert, ruhend und entspannt, lässig und relaxt. Der Schwung des Rückens und das Volumen des Sitzes stehen in ungewöhnlicher Spannung zueinander. Der Rücken bietet zugleich eine bequeme Armauflage.

Das Design zeichnet sich durch elegante Kurven und schwungvolle Formen aus, die die Energie unterstreichen, die in dem Sessel steckt. Den »Drift« Sessel gibt es in zwei Versionen, nach links oder nach rechts spiegelbildlich ausgeformt.

This armchair is a striking soloist. Fitted with a swivel support, »Drift« very easily becomes a »motion chair«. One can swing round to face an interlocutor, sit dynamically with slight movements to the left or right, or simply change one's line of sight from time to time.

Moreover, the organic cushions allow a wide variety of sitting postures: upright and concentrated, restfully at ease, casual and relaxed. There is an unusual tension in the relationship between the sweep of the back and the bulk of the seat. The back doubles as a comfortable armrest.

Distinctive features of the design are the elegant curves and sweeping shapes that emphasize the energy in the chair. The »Drift« armchair comes in two versions that mirror each other, facing either to the left or the right.

Jury
Der Sessel überzeugt durch seine Größe, seine Funktionalität und seine Beweglichkeit. Außergewöhnlich das mobile Sitzen: Man kann sich in alle Richtungen drehen und mit allen Leuten unterhalten. Sehr kommunikativ, schöne organische Form, spielerische Aspekte. Sogar als Zweisitzer geeignet, auch wenn er dafür etwas klein ist. Überzeugendes Design. Der Energie-Aspekt: er bewegt sich.
—
Judges panel
An impressive armchair—due to its size, its functionality and its mobility. Mobility while seated is an unusual feature: one can turn in any direction to talk to anyone in the room. Very communicative, attractive organic form, playful aspects. Can be used as a two-seater, even if a little too small. Impressive design. The energy angle: it moves.
—

Circle
Sofa / Sofa

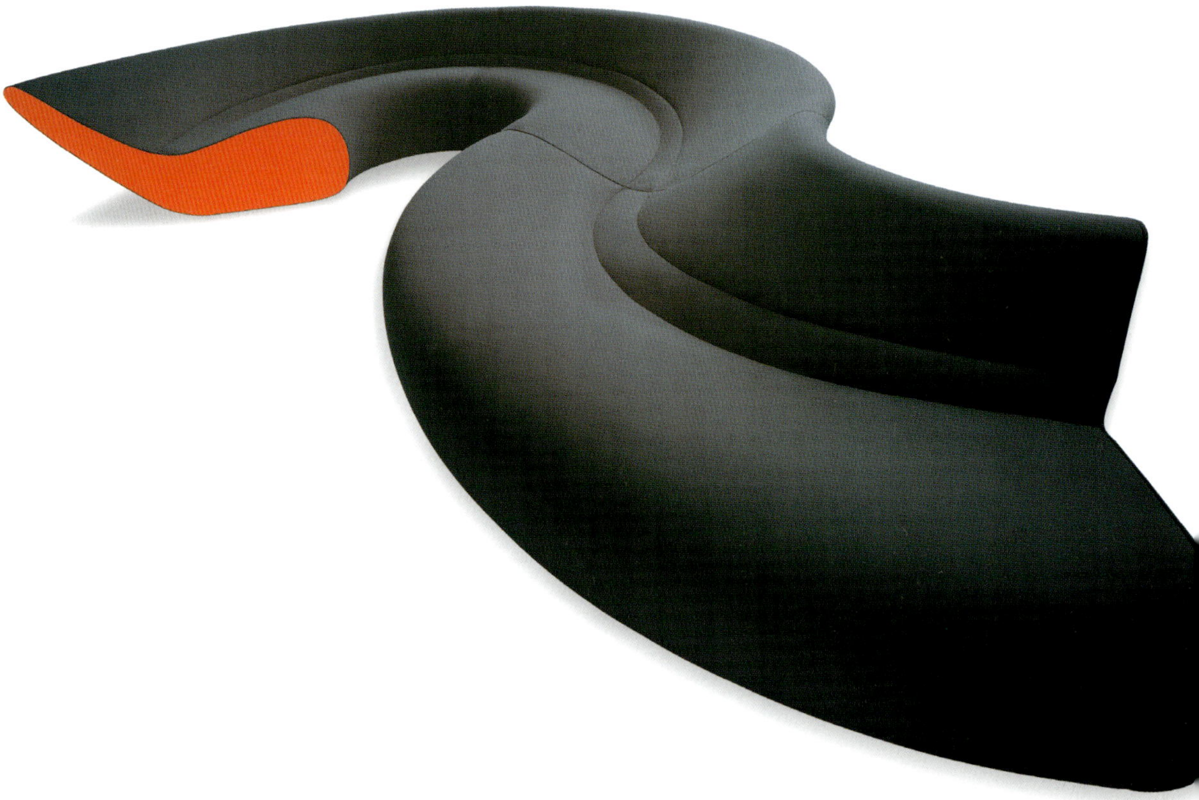

Circle

Sofa

Hersteller / Manufacturer
Walter Knoll AG & Co. KG
D-71083 Herrenberg
—
Design / Designer
UN Studio
Ben van Berkel
NL-1073 AX Amsterdam
—
Vertrieb / Distributor
Walter Knoll AG & Co. KG
D-71083 Herrenberg
—

»Circle« entstand im Zusammenhang mit dem Entwurf für das Mercedes-Benz-Museum in Stutgart, dessen Konzept die Designidee anregte. Aus der Verbindung von konvexen und konkaven Formen entstehen dabei schwungvolle Sitzskulpturen, die eigene Räume bilden.

»Circle« besteht aus vier Teilen, je zwei bilden einen Halbkreis, drei eine kreisrunde Plastik, die zum Begehen geöffnet ist und alle vier einen geschlossenen Kreis. »Circle« kann sich aber auch wie ein Reptil durch den Raum schlängeln. Die Neigung des Rückens und die Sitztiefe verändern sich dabei fließend vom aufrechten Sitzen bis zum entspannten Liegen. Konstant bleibt allein die Höhe.

Die Nutzungsmöglichkeiten reichen vom Relaxen, auf dem Sofa Lümmeln, über kurzes Verweilen in Wartesituationen bis hin zu angeregten Gesprächsrunden. Das Design besticht durch kraftvolle organische Formen und die Dynamik der Linien und Kurven.

»Circle« was developed in connection with the design for the Mercedes-Benz Museum in Stuttgart. The museum concept engendered the design idea. Sweeping seat sculptures forming spaces in their own right result from the combination of convex and concave shapes.

»Circle« comprises four parts. Two parts form a semicircle, three an open circular sculpture that can be entered, and all four a closed circle. However, »Circle« can also resemble a reptile, snaking through a room. The inclination of the backrest and the seat depth change continuously, from sitting upright to relaxed recumbency. Only the height is constant.

Usage options range from relaxing or lounging on the sofa through short periods of waiting to animated discussions in groups. A captivating design, thanks to powerful, organic forms and the dynamics of the lines and curves.

Jury
Die flexible dynamische Sitzlandschaft hat mehrere Funktionen. Zum einen ist sie als Wartesituation im öffentlichen oder halböffentlichen Bereich einzusetzen, zum anderen besitzt sie durch das Polster eine große Gemütlichkeit und kann auch als absolutes Relaxmöbel beispielsweise in einer Lounge oder einer Bar genutzt werden.
—
Judges panel
An adaptable, dynamic seating landscape with several functions. On the one hand, it can be used as a waiting facility in public or semi-public buildings; on the other, the cushions make it extraordinarily comfortable, so that it can also be used for total relaxation, e.g. in a lounge or bar.
—

May, August, Julies
Stauraummöbel / Storage furniture

May, August, Julies

Stauraummöbel
Storage furniture

Hersteller / Manufacturer
Leise®
D-60311 Frankfurt am Main
—
Design / Designer
Leise Studio
Knut Völzke
D-60311 Frankfurt am Main
—
Vertrieb / Distributor
Leise®
D-60311 Frankfurt am Main
—

»May, August, Julies« sind Objekte zum Benutzen, die nicht auf eine bestimmte Funktion festgelegt sind. Sie sind vielseitig verwendbar, weil sie ihre Bestimmung offen lassen und mit der Abstraktion spielen. Sie sind Raumkörper und Stauräume zugleich.

Die Objekte, die es in sieben verschiedenen Farben gibt, zeichnen sich durch zurückhaltende Gestaltung und ein besonderes Material aus. Sie bestehen aus einer speziellen, mit farbigem Schichtstoff belegten, feinen Stäbchenplatte.

»Julies«, das sind Lowboard, Sofatisch, Wäschebox, Nachtkonsole, kleines Regal, Beistelltisch, Hocker oder Sitzbank. Kombiniert mit »May« und »MayMay«, die als Deckelelemente benutzt werden können, werden sie zu Aufbewahrungsmöbeln. »August« dagegen ist ein Solitär, ein kleiner Würfel als Ablage oder Podest, nützlich und doch abstrakt.

»May, August, Julies« are utility objects, but without any specific function. They can be used in many different ways, because they leave their purpose undefined and play with abstraction. They are simultaneously spatial objects and storage spaces.

The objects, available in seven colours, stand out by virtue of their discreet design and their special materials. They are made of a special fine laminboard covered with a coloured laminate.

»Julies« are lowboards, coffee tables, laundry boxes, bedside cabinets, small shelves, occasional tables, stools or benches. Combined with »May« and »MayMay«, which can be used as lids, they become storage units. By contrast, »August« is a loner, a small cube that can be used for storage as a rostrum, practical but nonetheless abstract.

Jury
Einfaches, nettes Systemmöbel, gut gestaltet, qualitativ hochwertig, flexibel und mobil, überlebt mehrere Umzüge. Überzeugend sowohl die Funktionalität als auch die Formensprache und die Verarbeitungsqualität. Neue Interpretation von Systemmöbeln.
—
Judges panel
Simple, nice system furniture, well designed, of high quality, versatile and movable, will survive several relocations. Not only their functionality, but also the formal expression and the workmanship are impressive. A reinterpretation of system furniture.
—

Dresscode

Schrank
Wardrobe

Hersteller / Manufacturer
Nils Holger Moormann GmbH
D-83229 Aschau im Chiemgau
—
Design / Designer
Jörg Boner
CH-8045 Zürich
—
Vertrieb / Distributor
Nils Holger Moormann GmbH
D-83229 Aschau im Chiemgau
—

Der schlichte, ganz reduziert gestaltete Schrank aus ungewöhnlichem Material besteht aus wenigen Einzelteilen und ist in ein paar Minuten aufgebaut. Dank seines minimalen Transportmaßes und seines geringen Gewichts ist er leicht zu befördern. Er sieht nicht nur als Einzelstück gut aus sondern auch mehrere Schränke aneinandergereiht oder als Raumteiler.

Die Wände sind in einem Rahmen aus unbehandeltem Eschenholz fixiert und bestehen aus einem neuartigen Textilverbundmaterial, das im Thermo-Verfahren zu stabilen Bahnen verpresst ist. Das Innenleben lässt sich individuell mit Fachböden und Kleiderstangen ausgestalten.

»Dresscode« ist in verschiedenen Farben und zwei Breiten erhältlich. Die Innenseite ist schwarz, die Außenseite gibt es wahlweise in rot, olivgrün oder grau.

Jury
Sehr einfache leichte Lösung, schlicht und angemessen aus einheimischen Materialien und einheimischer Produktion. Neue Formensprache, unaufdringlich. Interessantes Gestaltungselement sind die über Kreuz geführten Seitenpaneele.
—
Judges panel
Very simple, lightweight solution, unembellished and appropriate, locally made from local materials. New formal expression, discreet. The crossed-over side panels are an interesting design feature.
—

A simple wardrobe with an extremely reduced design, made of an unusual material. It comprises just a few components and takes only a few minutes to assemble. Thanks to its minimal transport dimensions and low weight it is easy to transport. It looks good not only alone but also when several are placed in line or used as a room divider.

The side panels are fixed in a frame of untreated ash. They are made of a new type of textile composite that is thermo-pressed into robust webs. The interior can be fitted out individually with shelves and clothes racks.

»Dresscode« is available in various colours and two widths. The interior is black, the exterior is available in red, olive green or grey.

Trick Stick
Garderobe / Coat stand

Trick Stick

Garderobe
Coat stand

Hersteller / Manufacturer
Nils Holger Moormann GmbH
D-83229 Aschau im Chiemgau
—
Design / Designer
Patrick Frey + Markus Boge
D-30451 Hannover
—
Vertrieb / Distributor
Nils Holger Moormann GmbH
D-83229 Aschau im Chiemgau
—

Diese außergewöhnliche Garderobe mit Objektcharakter besteht lediglich aus drei Eschestäben, die nach traditioneller japanischer Verbindungstechnik ohne Werkzeug zusammengesteckt werden. Zudem hat man beim Aufbau von »Trick Stick« Spaß am Knobeln wie bei einem Puzzle. Zur Sicherheit wird jedoch die Anleitung mitgeliefert.

Braucht man »Trick Stick«, steht er auf einem Bein, lehnt an der Wand und dient als Ablage für Hüte, Schals und Jacken. Wird er nicht benötigt, können die drei Stäbe platzsparend aufbewahrt werden. Die drei Stabenden sind in den Farben weiß, rot und blau lackiert. Ein Rutschstop aus Silikon sorgt dafür, dass »Trick Stick« fest steht.

This unusual coat stand, with the allure of an objet d'art, consists merely of three sticks of ash connected without tools, by a traditional Japanese method. Assembling »Trick Stick« is fun—it is a kind of puzzle. However, to be on the safe side assembly instructions are supplied.

When »Trick Stick« is needed it stands on one leg, leaning against the wall, as a place to hang hats, scarves and jackets. When it is not needed the three sticks can be disassembled to save space. The ends of the sticks are painted white, red and blue. A silicone antislip foot ensures that »Trick Stick« stands firm.

Jury
Witziges Möbel, wie ein Puzzle, spricht den Spieltrieb an. Das Demontieren und Montieren ist Teil des Objekts. Dass das schwierig ist, ist Absicht der Gestalter. Es kostet Energie und Nerven. Wunderbar, dass man keine Schrauben braucht. Sympathisch, dass es lediglich eine Konstruktion ist »ohne Spuren von Design«. Sehr schön das Material und die Dimensionen. Bezieht sich auf Rietveld, der die gleiche Art von Konstruktion und Farbcodes entwarf.
—
Judges panel
An amusing piece of furniture, like a puzzle, appeals to the play instinct. Assembly and disassembly is an integral part of it. It is difficult to assemble and that is the designer's intention. It saps energy and tries the nerves. Wonderful that no screws are needed. Pleasant that it is merely a construction, »without any trace of design«. The material and the dimensions are very agreeable. Harks back to Rietveld, who designed the same type of constructions and colour codes.
—

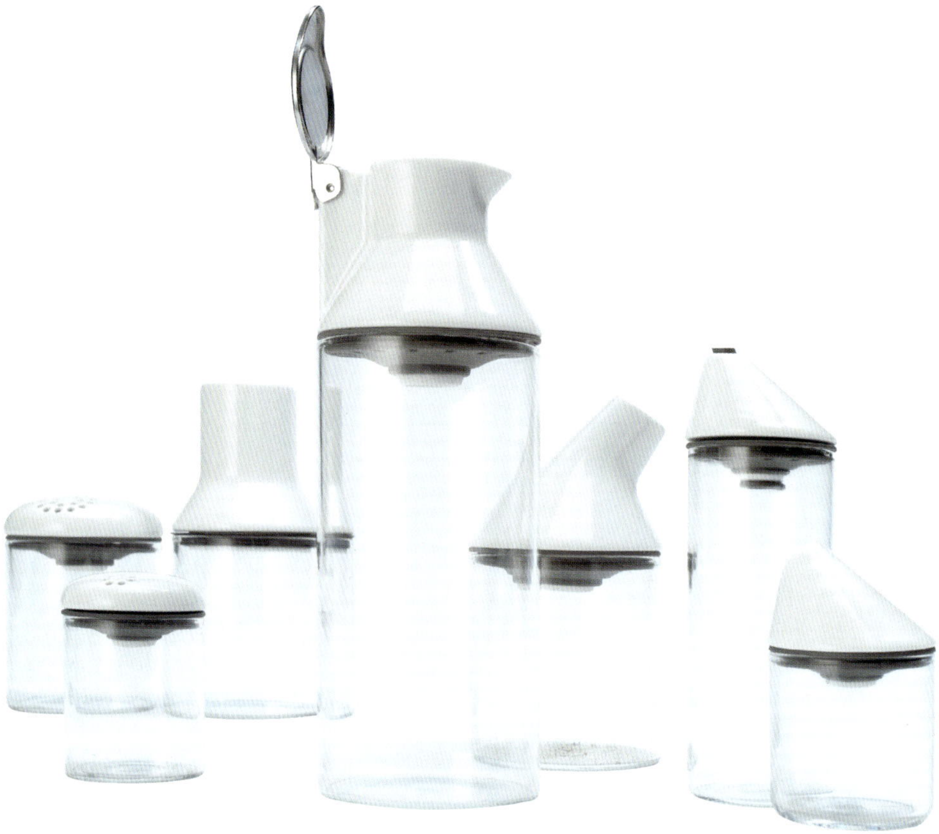

Shuttle
Aufbewahrungssystem / Storage system

Shuttle

Aufbewahrungssystem
Storage system

Hersteller / Manufacturer
Thomas – The Trend Factory
Rosenthal AG
D-95100 Selb
–
Design / Designer
Stefan Diez
D-80469 München
–

»Shuttle« ist ein außergewöhnliches Aufbewahrungssystem in futuristischem Design, das Form und Funktion vereint. Es besteht aus einer Reihe von Glascontainern in fünf verschiedenen Durchmessern und mit unterschiedlichen Aufsätzen wahlweise aus Kunststoff oder Porzellan, die sich vielfältig kombinieren und verwenden lassen.

So geben beispielsweise hochgezogene Porzellandeckel in drei Größen Marmeladen- oder Zuckerlöffeln den nötigen Halt. Ein Porzellanaufsatz für Essig- und Ölgefäße besitzt im Innern ein Kunststoffröhrchen, das mögliche lästige Flecken auf dem Tischtuch vermeidet. Aufsätze für Streuer, ob für Zucker, Gewürze oder Kakao, runden das Programm ab. Neuartige flache Kunststoffverschlüsse schließen luftdicht und sorgen dafür, dass das Aroma erhalten bleibt.

Zu dem Aufbewahrungssystem gehört eine ungewöhnliche asymmetrische Zitruspresse, die durch ihre einfache Handhabung besticht. Die Glasunterteile bestehen aus Borosilikat, einem hitzebeständigen, schockfesten Glas, das auch für die Mikrowelle geeignet ist.

»Shuttle« is an unusual, futuristically designed storage system that unites form and function. It comprises a series of glass containers, each of a different diameter and with different lids or caps. These are available in plastic or porcelain, and can be combined and used in a variety of ways.

Elongated porcelain caps, for example, available in three sizes, provide the support needed for jam or sugar spoons. A porcelain cap for oil and vinegar bottles has an in-built plastic tube that prevents any irksome stains on the table cloth. The range is rounded off by dispenser tops—for sugar, spices or cocoa. Innovative, flat plastic lids provide an airtight seal for the containers, ensuring that the aroma stays in.

The storage range also includes an unusual, asymmetrically shaped lemon squeezer, which is strikingly simple to handle. The glass containers themselves are made of borosilicate, a heat-resistant, shockproof glass that is also suitable for the microwave.

Jury
Bestechend ist vor allem die Zitruspresse. Einfaches Design, leicht zu reinigen. Sehr ergonomische Lösung für die Bewegung, die man ausführen muss, um eine Orange zu pressen, eine konzentrische Kreisbewegung. Überzeugend auch das Design der Deckel und des Rings als Zitat einer Coladose, ebenso wie die klare Formensprache der Glasbehälter. Sehen sowohl in der Küche als auch auf dem Tisch gut aus.
–

Judges panel
The lemon squeezer is especially impressive. Simple design, easy to clean. Very ergonomic solution for the movement you have to make to squeeze an orange—a concentric circular movement. The lid design also makes sense, with its ring as a nod to the coke tin, as does the clear formal expression of the glass containers. Look good in the kitchen and on the dining table.
–

Opening

Vasen
Vases

Hersteller / Manufacturer
Rosenthal Studio-Line
Rosenthal AG
D-95100 Selb
—
Design / Designer
Little Wonder
Gyungju Chyon und John Stanislav Sadar
c/o Rosenthal Studio-Line
Rosenthal AG
D-95100 Selb
—

Auf den ersten Blick wirken die Vasen »Opening« bekannt. Die Form erinnert zunächst an typische Keramikformen, wie sie beispielsweise schon die Minoer herstellten.

Die Öffnung jedoch ist außergewöhnlich und faszinierend, sie bricht mit der Tradition. Sie ist der Natur, die die schönsten Formen hervorbringt, entliehen und erinnert an aufgebrochene Eierschalen.

Asiatische Einflüsse und hohe Kunstfertigkeit vereinen sich bei den Vasen, deren Schönheit auch ohne Inhalt voll zur Geltung kommt. Sie wirken eher wie Porzellan-Objekte. »Opening« gibt es in verschiedenen Größen und Formen.

At first glance, the »Opening« vases seem familiar. Their form is initially reminiscent of typical ceramic forms produced by the Minoans.

Their opening, however, is unusual and fascinating, and breaks with the above tradition. It is borrowed from the world of nature, which produces the most beautiful forms, and recalls the form of a boiled egg that has been opened.

Combining Asian influences and outstanding craftsmanship, the vases are equally appealing even when empty. Their effect is more that of a porcelain objet d'art. »Opening« is available in various shapes and sizes.

Jury
Dynamisch explodierende Form, materialgerecht, die Zufallsform der Öffnung überzeugt. Hier explodiert etwas mit viel Energie, die die perfekte Form durchschlägt. Dieses ästhetische Element passt sehr gut zu dem Wettbewerbsthema.
—
Judges panel
Dynamically explosive form, appropriate use of material, the aleatory shape of the opening is nicely done. Here we have an object exploding with surplus energy that bursts out of the vases' perfect form. The aesthetic element is extremely apposite, given the competition theme.
—

Lounge

Schalen
Bowls

Hersteller / Manufacturer
WMF Württembergische Metallwarenfabrik AG
D-73312 Geislingen
—
Design / Designer
James Irvine
I-20129 Mailand
—
Vertrieb / Distributor
WMF Württembergische Metallwarenfabrik AG
D-73312 Geislingen
—

Die Schalen der Serie »Lounge« setzen dekorative Akzente und schaffen eine individuelle Wohnatmosphäre. Ihre auf das Wesentliche reduzierte Gestaltung vermittelt Ruhe und Klarheit. Sie bestechen durch großzügige Radien und puristische, organische Formen mit etwas üppiger Tendenz. Ihre doppelwandige Verarbeitung ermöglicht den Eindruck einer massiven Wandstärke.

»Lounge« ist in perfekter Handwerklichkeit aus mattem Edelstahl gefertigt. Die Vasen gibt es in fünf verschiedenen Größen. Platzsparend ineinander gestapelt werden sie zu einem außergewöhnlichen Designobjekt und einem attraktiven Blickfang.

Auch in ihrer Verwendung sind sie höchst vielseitig. So dienen sie als dekorative Schalen für Obst, feine Plätzchen oder salzige Knabbereien. Aber sie sind auch perfekt für Schwimmkerzen, getrocknete Blüten, Zierkürbisse und vieles mehr.

Jury
Die Schalen im Wohnbereich wirken sowohl ineinandergestapelt als auch als Einzelschalen in sich als geschlossene Form und passen dank ihrem hochwertigen Material in ganz verschiedene Ambientes.
—
Judges panel
Both when stacked together and on their own, these bowls give the impression of a perfect form in the living room. Thanks to their high-quality material, they go well with a wide variety of living environments.
—

The bowls of the »Lounge« series set decorative accents and create an individual living atmosphere. Reduced to the bare essentials, their design conveys a sense of tranquillity and clarity. They stand out by virtue of their generous dimensions, and purist, organic forms that tend towards opulence. Because of their double-walled design, their appearance is one of solidity.

»Lounge« is perfectly crafted out of matt stainless steel. The vases are available in five different sizes. Stacked one inside the other to save space, they take on the form of an unusual design object and attractive eye-catcher.

In terms of use, they are also extremely versatile. They can be used as ornamental bowls for fruit, biscuits or savouries. But they are also perfect for floating candles, dried flowers, decorative squashes and much more.

Jury

Interessante Idee, die Feder, die normalerweise innen liegt, außen zu benutzen und als einziges zusätzliches Element zu einer Standardmine zu verwenden. Hier wurde mit wenig Materialeinsatz ein interessanter Kugelschreiber entworfen. Ideal für kurze Notizen und Ähnliches.

—

Judges panel

Interesting idea—the spring that is normally inside the pen is now on the outside. The only additional element is a standard refill. With a bare minimum of materials, a very interesting ballpoint pen has been designed here. Ideal for taking brief notes and the like.

—

»On« Mine nach unten schieben, bis sie einrastet. (1–3)

»Off« Auf das Rahmenende drücken, bis die Mine aus ihrer Halterung rutscht. (4–6)

—

»On« Apply pressure to the refill until it clicks into place. (fig. 1–3)

»Off« Apply pressure to the end of the frame until the refill slips out of its fixed position. (fig. 4–6)

—

1

2

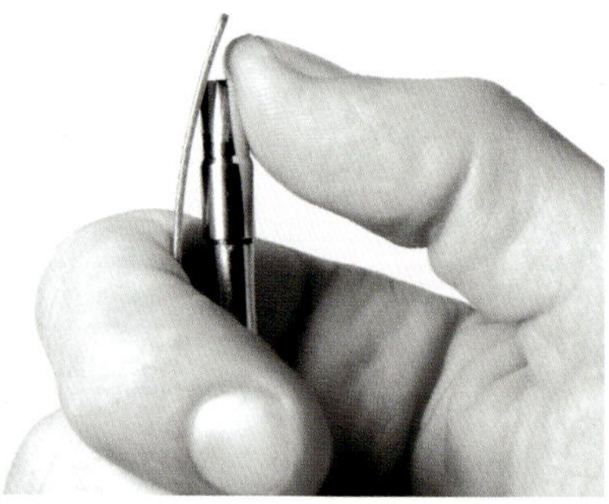

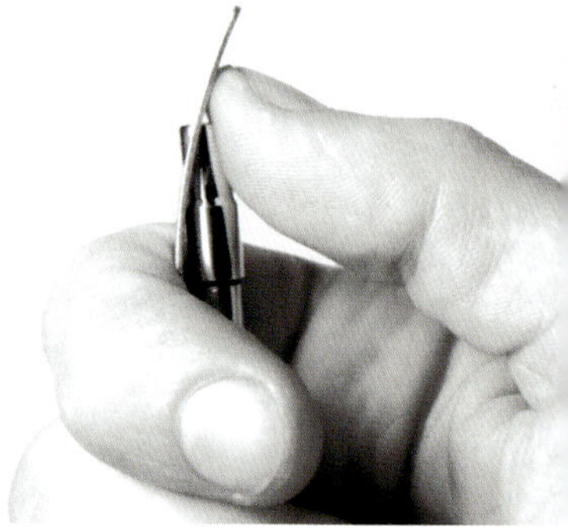

4

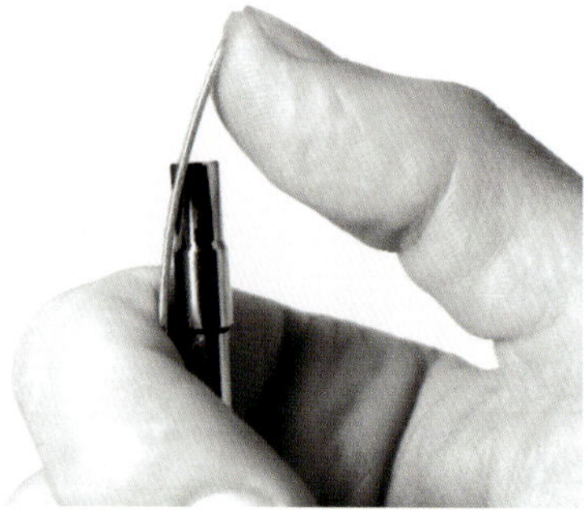

5

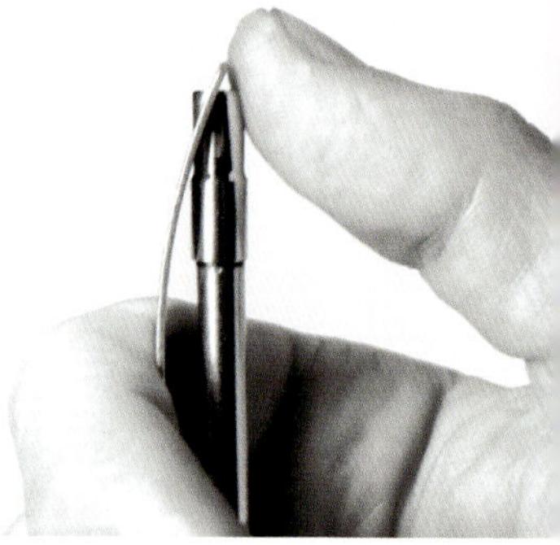

Patent 1

Kugelschreiber
Ballpoint pen

Hersteller / Manufacturer
Albert Ebenbichler
D-73728 Esslingen
—
Design / Designer
Werksdesign / In-house design
Albert Ebenbichler
—
Vertrieb / Distributor
Albert Ebenbichler Innovation
D-73728 Esslingen
—

Wenige Produkte werden so schnell zu Wegwerfprodukten wie ein Kugelschreiber. Und er besteht häufig aus zahlreichen Einzelteilen. Nicht so der »Patent 1«.

Der Kugelschreiber besteht lediglich aus einem präzise geschwungenen Federstahl, der gemeinsam mit einem Griffstück und einer handelsüblichen Mine zu einer exakt funktionierenden Einheit verschmilzt. Bei ihm werden Form und Funktion eins.

Allein der schlichte, sanft nachgebende Metallrahmen erfüllt alle Funktionen. Er lässt die Mine mühelos in seinem offenen Inneren versinken und wieder zum Vorschein kommen. Um den Kugelschreiber in Schreibstellung zu bringen, wird die Mine mit sanftem Druck nach unten geschoben bis sie einrastet. Drückt man auf das Ende des Rahmens, bis die Mine aus ihrer Halterung rutscht, schnappt sie in ihre Grundposition zurück.

Der dynamische Schwung verbunden mit der technischen Schlichtheit weckt den Spieltrieb. Die Standardmine ist mit einem Adapterstück versehen und mit einer einzigen Bewegung zu wechseln. »Patent 1« lässt sich auch bequem an die Jackentasche stecken.

Few products are thrown away as quickly as ballpoint pens. Frequently, these pens are made up of a number of individual parts. But not »Patent 1«.

This ballpoint pen is nothing but a precisely curved steel spring, merged with a grip and a commercially available refill to form a precisely working unity. Form and function are unified in this pen.

The unassuming, slightly elastic metal frame fulfils every function on its own. It allows the refill to disappear easily into its open interior, and reappear again just as easily. To prime the ballpoint pen, gentle pressure is applied to the refill until it clicks into place. By applying pressure to the end of the frame, the refill can be made to slip out of its fixed position and return to its original position.

This combination of dynamic energy and technical simplicity awakens the user's sense of play. The standard refill is fitted with an adapter, and can be replaced in one single movement. »Patent 1« can also be easily carried in a jacket pocket.

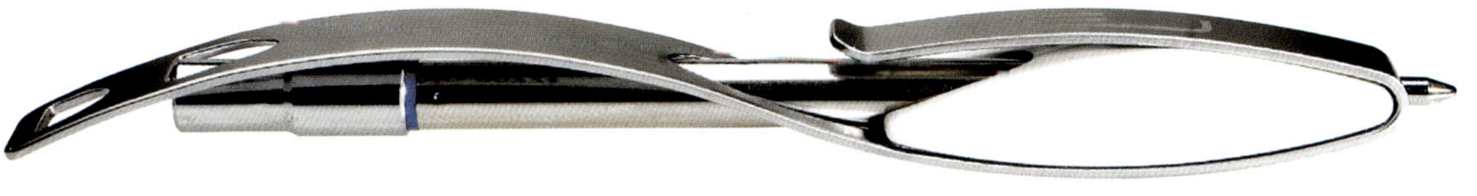

Ambiente / Silber
Ambience / Silver

Lamy Studio

Füllhalter
Fountain pen

Hersteller / Manufacturer
**C. Josef Lamy GmbH
D-69123 Heidelberg**
–
Design / Designer
**Werksdesign / In-house design
und Hannes Wettstein
Zed AG
CH-8008 Zürich**
–
Vertrieb / Distributor
**C. Josef Lamy GmbH
D-69123 Heidelberg**
–

»Lamy Studio« ist ein Schreibgeräteprogramm, zu dem neben Drehkugelschreibern drei verschiedene Füllhalter gehören. Zwei davon besitzen eine Stahlfeder, die Oberfläche des einen ist aus strichmattiertem Edelstahl, des anderen aus mattschwarzem Softlack. Das dritte Modell für hohe Ansprüche besitzt ein edles Palladium-Finish und eine rhodinierte 14-kt-Goldfeder mit Goldintarsie.

Besonders die außergewöhnliche Gestaltung des Clips sticht hervor. Dieser glanzverchromte Clip hat trotz hoher Funktionalität nichts mehr mit einem normalen Schreibgeräteclip zu tun. Seine elegant in sich gedrehte Form erinnert eher an eine moderne Plastik etwa von Constantin Brancusi oder Max Bill. Weitere markante Gestaltungselemente wie die ebenfalls glanzverchromten Geräteenden machen ihn zusätzlich zu einem Blickfang.

Der »Lamy Studio« liegt sehr gut in der Hand. Sein großer Durchmesser sorgt für überdurchschnittlich hohen Schreibkomfort.

»Lamy Studio« is the name of a range of writing implements including ballpoint pens and three different fountain pens. Two of them come with a steel nib, but while the surface of the one is brush-finished stainless steel, the other is available in black soft finish. The third model, for more discerning tastes, comes in a fine palladium finish and a rhodanized 14-carat gold nib with a gold inlay.

The unusual design of the clip is especially striking. With its shiny chrome finish, the clip is completely unlike any other, despite its very functional character. Its elegant, twisted form is reminiscent of a modern sculpture by Constantin Brancusi, for example, or Max Bill. Other striking design elements, such as the end pieces on the cap and barrel, again in a shiny chrome finish, also contribute to its eye-catching quality.

»Lamy Studio« fits very well in the hand. Its large diameter makes it extremely pleasant to write with.

Jury
Lamy investiert seit vielen Generationen von Schreibgeräten sehr viel Intelligenz in die Gestaltung der Clips. Auch hier hervorragende Lösung: Der Clip lässt sich durch den großen Radius, der auch formal gut zum Füllhalter paßt, sehr gut an Jackentaschen stecken. Das Palladium hat eine haptisch sehr ansprechende Oberfläche. Langlebiges Design durch starken emotionalen energetischen Wert.
–

Judges panel
For many generations of writing implements, Lamy has invested a lot of intelligence in the design of its clips. Here again, we have an excellent solution: Because of its large radius, which also complements the fountain pen formally, it can very easily be inserted into a jacket pocket. The tactile quality of the palladium is very appealing. Its strong energetic emotionality means this is a design that will keep its appeal for a long time.
–

Base Kid
Schreiblernfüller / Beginner's fountain pen

Base Kid

Schreiblernfüller
Beginner's fountain pen

Hersteller / Manufacturer
Schneider Schreibgeräte GmbH
D-78144 Tennenbronn
–
Design / Designer
Atelier für Gestaltung Papenfuss
D-99423 Weimar
–
Vertrieb / Distributor
Schneider Novus Vertriebs GmbH
D-82362 Weilheim
–

Der Schreiblernfüller »Base Kid« berücksichtigt nicht nur den heutigen Schreibunterricht, der stärker von Bewegung geprägt ist, sondern er wächst auch mit. Dafür sorgt ein Baukastensystem, bei dem die einzelnen Teile und Farben miteinander kombiniert und zahlreiche individuelle Füller und Patronenroller zusammengestellt werden können.

Das ergonomisch geformte Griffstück für Schulanfänger ist elastisch und hat eine Oberfläche ohne störende Rillen oder Noppen. Jeder Finger findet sicher seinen Platz auf großen Auflagenflächen, so dass der Druck optimal verteilt wird. Wenn die Hand größer wird, kann der kurze Schaft gegen einen langen getauscht werden. Ausgeformte Abschlüsse am Füller und an der großen Kappe, die zu spielerischen Entspannungsübungen anregen, sichern gegen Wegrollen.

Der transparente Schaft gibt den Blick auf den Tintenvorrat frei. Alle Gehäuseteile sind aus dickwandigem Kunststoff hergestellt und dadurch besonders robust. Für die Bedürfnisse jugendlicher Schreiber werden den Basisteilen wenige Teile und neue Materialien, die für »coole« Schreibgeräte sorgen, hinzugefügt.

The »Base Kid« beginner's fountain pen not only takes account of modern teaching methods for handwriting, which involve much more movement, but also grows with the child. This is ensured by the modular system, whose individual parts and colours can be combined with each other, allowing several individual fountain pens and felt pens to be put together.

The ergonomically formed grip for school beginners is elastic, and its surface is uninterrupted by grooves or pimples. Each finger finds a secure grip on the large grip surface, allowing optimum distribution of pressure. As the child's hand grows, the short shaft can be exchanged for a long one. Specially moulded end pieces on the barrel and the large cap not only encourage playful relaxation, but also prevent them from rolling away.

The transparent shaft means the child can see how much ink is in the pen. All the housing parts are made of thick plastic, making them especially sturdy. For older schoolchildren, a few parts and new materials can be added to the pen's basic elements, and make for a »cool« writing implement.

Jury
Interessante Aspekte: zum einen dieser modulare Aufbau und die Produktphilosophie, die sehr effizient ist, zum anderen die Formensprache, die sich wirklich an die richtige Zielgruppe richtet, nämlich Kinder und Eltern. Sehr gefälliges Design, es bringt Kindern bei, aufzupassen, wenn sie mit dem Füllfederhalter schreiben.
–
Judges panel
Interesting features: on the one hand, modular structure and a very efficient product philosophy, and on the other hand a formal expression that truly appeals to the target group—children and their parents. Very appealing design, it teaches children to pay attention when they write with a fountain pen.
–

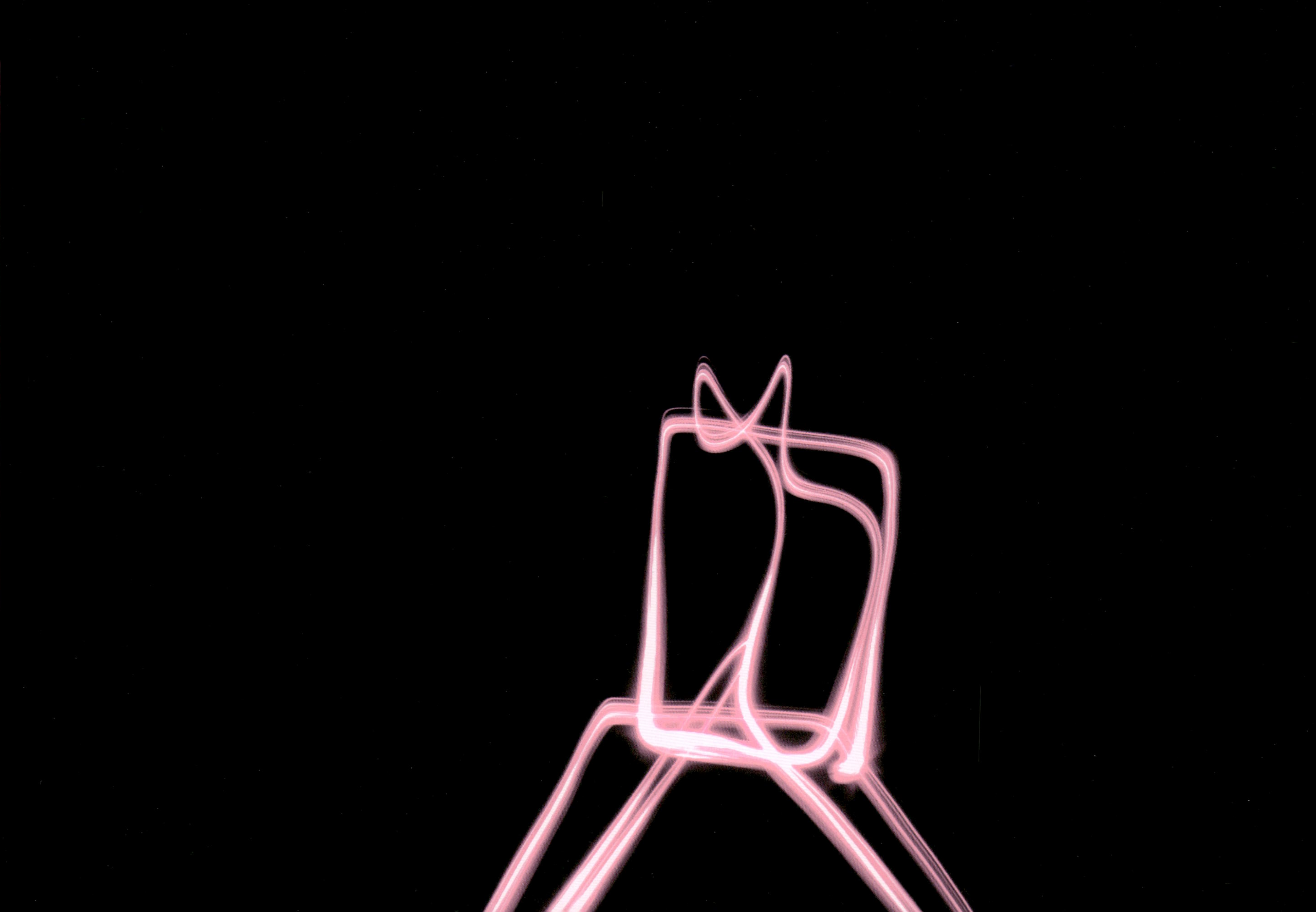

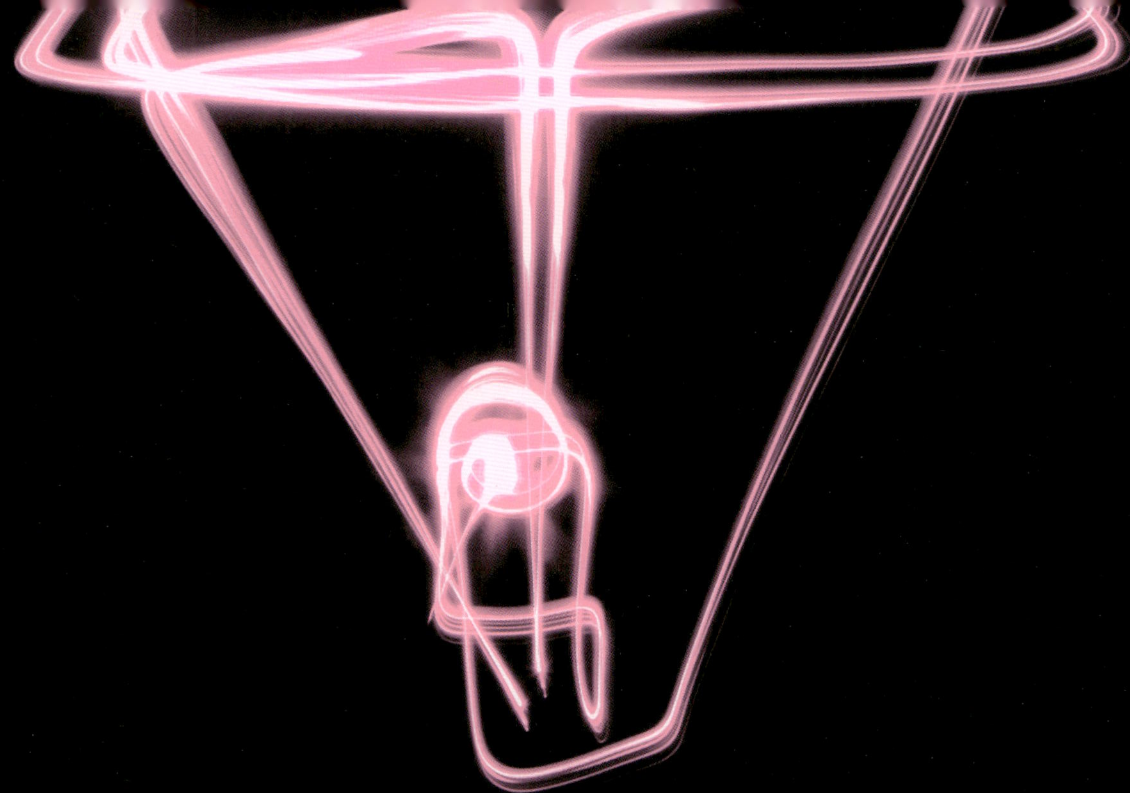

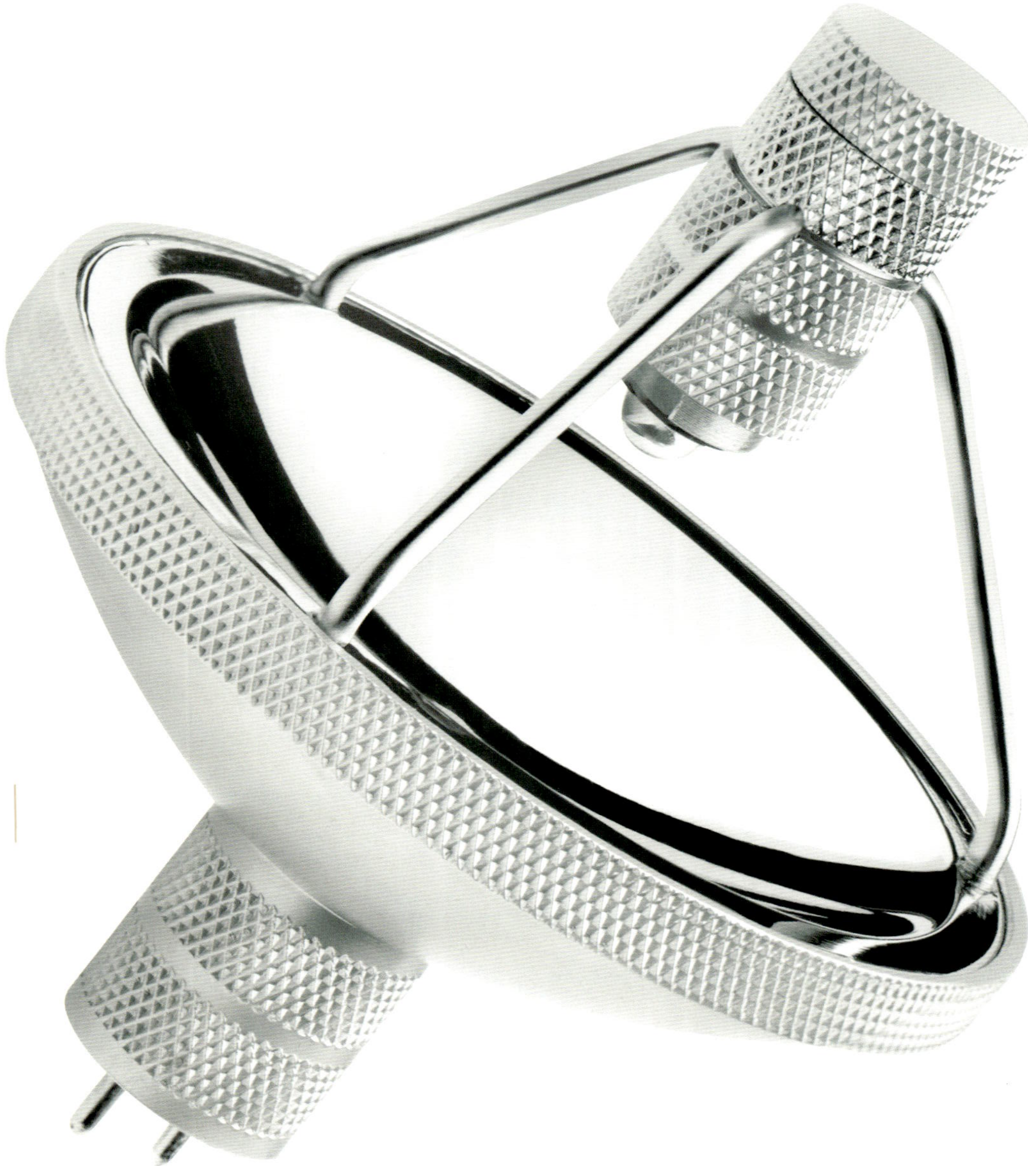

Parabolico

LED Strahler
LED lenser

Hersteller / Manufacturer
Zweibrüder Optoelectronics GmbH
D-42699 Solingen
—
Design / Designer
Werksdesign / In-house design
Rainer Opolka
—
Vertrieb / Distributor
Zweibrüder Optoelectronics GmbH
D-42699 Solingen
—

Der LED Strahler »Parabolico« ist der erste fokussierbare 12 Volt Strahler. Durch Drehen an der Rändelschraube ist der Lichtkegel über den Abstrahlwinkel des patentierten Parabolspiegels mit einem Durchmesser von 65 Millimetern individuell einstellbar.

Der »Parabolico« gehört zu einer Serie von Reflektoren, die auch in Helligkeit, Lebensdauer, Umweltfreundlichkeit und Stoßfestigkeit anderen Strahlern weit überlegen sind. Mit ihnen wurden elegante und beispiellos helle Spots entwickelt, die Aufsehen erregen und durch das schlichte Design gefallen. Sie setzen nicht zuletzt durch ihre Eigenschaften neue Maßstäbe, der Einsatz erhöht die Wohnqualität. Sie heben sich deutlich von normalen Halogenstrahlern ab und sind zugleich Energie sparende Gebrauchsgegenstände.

Durch den Einsatz modernster patentierter LED-Technik ist es möglich, fünf bis zehn Watt Halogenlicht durch diese Spots zu ersetzen, denn sie sind in der Form baugleich mit handelsüblichen zwölf Volt Halogenstrahlern. Der Einbau ist kein Problem. Das geht mühelos in alle gängigen zwölf Volt Beleuchtungssystemen mit GU 5.3 Fassungen: einfach den alten Spot herausziehen und durch die neuen LED Strahler ersetzen.

The »Parabolico« is the first-ever focusable 12-volt LED lenser. By turning the knurled screw, the beam can be individually adjusted via the radiation angle of the patented 65 mm diameter parabolic reflector.

The »Parabolico« is one of a series of reflectors that are far superior to other lensers in terms of brightness, service life, environmental compatibility and shock resistance. They are elegant, sensationally bright spots with a pleasantly uncluttered design. It is not least thanks to these characteristics that they set new standards of comfort in the home. They are a far cry from ordinary halogen spots, while at the same time being energy-saving requisites.

Thanks to state-of-the-art, patented LED technology these lensers can replace five to ten watts of halogen light, because their shape is identical with that of ordinary 12-volt halogen spots. They are easy to install in any standard 12-volt lighting system with GU 5.3 lamp-holders: you simply pull out the old spot and replace it with the new LED lenser.

Jury
Ungewöhnlich der niedrige Energieverbrauch bei zugleich sehr hoher Leuchtkraft. Das Licht reflektiert in einem Spiegel, einem parabolischen Reflektor. So bekommt man wesentlich mehr Lichtenergie von einer sehr viel kleineren Lichtquelle als bei herkömmlichen Spots. Sehr intelligente Kombination von alter und neuer Technologie. Super. Gutes außergewöhnliches Design.
—
Judges panel
Unusually low energy consumption for such high luminosity. The light is reflected in a mirror, a parabolic reflector. As a result, much more light is produced by a far smaller light source than conventional spots. Very intelligent combination of old and new technology. Great. Good, unusual design.
—

LED-Reflektor 12 Volt

Strahler
Lenser

Hersteller / Manufacturer
Zweibrüder Optoelectronics GmbH
D-42699 Solingen
—
Design / Designer
Werksdesign / In-house design
Rainer Opolka
—
Vertrieb / Distributor
Zweibrüder Optoelectronics GmbH
D-42699 Solingen
—

Die Reflektoren sind extrem helle Spots, gefertigt in High-End-Qualität. Sie heben sich durch ihr hochwertiges Design und den enorm geringen Stromverbrauch von üblichen Halogenstrahlern deutlich ab. Die LED Strahler können überall eingesetzt werden, sowohl im Wohn- und Arbeitsbereich als auch hervorragend zur Objekt- und Vitrinenbeleuchtung.

Ihre Lebensdauer ist 50-mal höher als bei herkömmlichen Halogenspots, sie sind wartungsarm und verbrauchen maximal 1,3 Watt. Dieser Bruchteil des Verbrauchs von herkömmlichen Halogenstrahlern spart nachhaltig Kosten. So kostet bei täglich 12 Stunden Leuchtdauer der Betrieb des Reflektors deutlich weniger als einen Euro pro Jahr.

Die Reflektoren werden in drei unterschiedlichen Bauformen angeboten: Neben dem »Parabolico« gibt es die Standard-Ausführung mit dem 19 mm Reflektor in den Lichtfarben rot, blau und grün und einen weiteren 12 Volt Reflektor mit einem patentierten Doppellinsen-Reflektor als Moonspot in den Lichtfarben weiß und warm weiß.

Reflectors are extremely bright spots, manufactured in high-end quality. They differ distinctly from conventional halogen spots by their high-quality design and extremely low power consumption. The LED spots can be used anywhere, both in residential and working environments, and are ideal for lighting objects and display cases.

Their service life is 50 times higher than that of conventional halogen spots, they require little maintenance and their power consumption is extremely low—a maximum of 1.3 watts. This is a fraction of the consumption of conventional halogen spots, and thus permanently saves money. Given 12 hours lighting a day, for example, lighting with the reflector costs well under one euro a year.

The reflectors are available in three different designs: besides the »Parabolico« there is the standard version with a 19 mm reflector and the light colours red, blue and green, as well as a 12-volt reflector with a patented double-lens reflector as a »Moonspot«, with the light colours white and warm white.

Jury
Schönes, klares Design, die Strahler sind vielseitig einsetzbar. Sehr gut die hohe Energieeinsparung im Vergleich zu herkömmlichen Halogenstrahlern, ihre Umweltfreundlichkeit und Langlebigkeit.
—
Judges panel
Attractive and clear design. The spots are versatile. Their high energy economy, eco-friendliness and long life as compared to conventional halogen spots are very good.
—

Beleuchtung / Silber
Lighting / Silver

LED-Lenser Automotiv
Aufladbare Autolampe / Rechargeable car light

LED-Lenser Automotiv

Aufladbare Autolampe
Rechargeable car light

Hersteller / Manufacturer
Zweibrüder Optoelectronics GmbH
D-42699 Solingen
—
Design / Designer
Werksdesign / In-house design
Wojtek Niemiec
—
Vertrieb / Distributor
Zweibrüder Optoelectronics GmbH
D-42699 Solingen
—

Diese kleine und ungewöhnlich helle All-round-Taschenlampe wird nicht, wie üblich, mit Batterien betrieben sondern mit wiederaufladbaren Akkubatterien. Aufgeladen wird sie einfach über den Zigarettenanzünder eines jeden Autos. Dank des leistungsstarken Nickel-Hydrid-Akku ohne Memory-Effekt ist das bis zu 1000-mal möglich.

Mit ihren gerade mal 4,8 Zentimetern Länge und einem Durchmesser von nur 2 Zentimetern ist die Lampe im Miniformat vielseitig einsetzbar: von der soliden Lampe in allen Belangen bis zur Damenlampe für die Handtasche. Sie ist nicht nur stoss- und schlagfest sondern auch wasserresistent. An- und ausgeschaltet wird sie einfach durch Drehen des kurzen Lampenkopfes.

Die LED-Lampe ist in Helligkeit, Lebensdauer, Stromverbrauch und Umweltfreundlichkeit jeder konventionellen Taschenlampe gleicher Größe weit überlegen. »Automotiv« besticht durch den kreisrunden bühnenspotartigen Lichtkegel, das hochwertige Gehäuse aus Edelstahl und das funktionale elegante und zeitlose Design.

This very small, unusually bright all-purpose flashlight is powered by rechargeable batteries. It is recharged simply by plugging it into a car cigarette lighter. Thanks to the high-performance nickel-hydride battery, which has no memory effect, up to 1000 recharges are possible.

Just 4.8 centimetres long and only 2 centimetres in diameter, this mini-torch is truly versatile—from a robust all-purpose flashlight to a light for a lady's handbag. It is not only shockproof, but also water-resistant. It is switched on and off simply by turning the short head.

With regard to brightness, service life, power consumption and environmental friendliness, this LED light is far superior to any conventional flashlight of the same size. The appeal of »Automotiv« lies in its perfectly circular beam, like a stage spotlight, the high-quality stainless steel housing and the functional, elegant, timeless design.

Jury
»Nettes kleines Spielzeug«, das sich im Zigarettenanzünder aufladen lässt. Überzeugend die LED-Technik und dass es sich an die Infrastruktur von Autos anpasst, man es aufladen kann und es dann immer griffbereit hat, wenn man eine Leuchte braucht.
—
Judges panel
Nice little toy that can be recharged in the cigarette lighter. The LED technology is impressive, also the fact that it adapts to the infrastructure of cars, that it can be recharged in the car and so is always to hand when a light is needed.
—

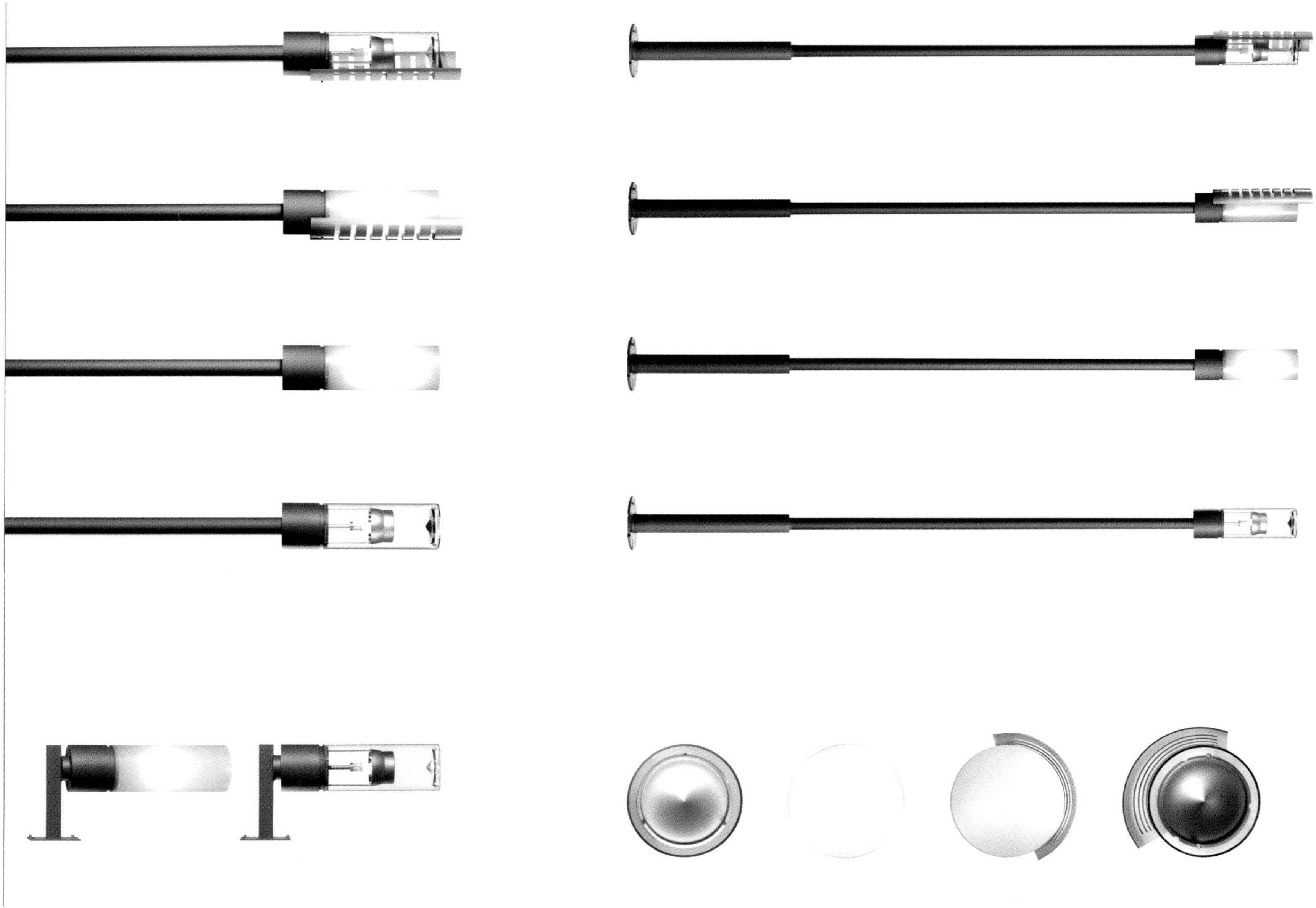

Residenza

Straßenleuchte
Street lamp

Hersteller / Manufacturer
Hess Form + Licht GmbH
D-78050 Villingen-Schwenningen
—
Design / Designer
Karsten Winkels
D-44147 Dortmund
—
Vertrieb / Distributor
Hess Form + Licht GmbH
D-78050 Villingen-Schwenningen
—

»Residenza« ist eine Leuchtenfamilie, die von Kommunen, Stadtwerken und Energieversorgern breit eingesetzt und für mehrstufige Lichtgestaltung angewendet werden kann. Sie eignet sich für die Straßenbeleuchtung in Wohn- und Mischgebieten, zur Beleuchtung von Fußgängerzonen und Plätzen oder um veraltete Pilz- und Kugelleuchten auf vorhandenen Masten zu ersetzen. Es gibt sie in ein- und zweifacher Ausführung sowie als Wandleuchte.

»Residenza« zeichnet sich durch moderne Lichttechnik, ein modulares Konzept, werkzeuglose Wartung und zurückhaltende, elegante Gestaltung aus. Dank dem variablen neuen Universalgehäuse ist die Montage auf verschieden dicken Standardmasten möglich.

Energie wird unter anderem eingespart, indem die nächtliche Leistung reduziert wird. Bei herkömmlichen Leuchten können bis zu 50 Prozent des Lampenlichtstroms ungerichtet verloren gehen und somit die Umwelt belasten. Bei »Residenza« dagegen wurde mittels modernster Reflektortechnologie das nach oben gerichtete Streulicht auf circa 5 Prozent des Lampenlichtstroms reduziert.

»Residenza« is a family of lamps that can be used in a wide range of situations by municipalities, borough service departments and energy suppliers, and for multiple-stage lighting design. It is suitable for street lighting in residential and mixed-use districts, for lighting pedestrian precincts and squares, or for replacing obsolete mushroom and ball lamps on existing lamp-posts. It is available in single and duplex versions, and as a wall-mounted lamp.

»Residenza« features modern lighting technology, a modular concept, tool-free maintenance and discreet, elegant design. Thanks to the new, variable universal housing it can be mounted on standard lamp-posts of different thicknesses.

Energy is saved by reducing the lamp's output at night. With conventional lamps, up to 50 per cent of the lamp's luminous flux may be undirected and thus wasted, polluting the environment. By contrast, in the case of »Residenza«, the amount of stray light directed upward has been reduced to about 5 per cent of the lamp's luminous flux.

Jury
Die Gestaltung der Leuchte mit allen Auflagen bezüglich Vandalismus-Sicherheit, Modularität und Nachrüstbarkeit ist gut gelöst. Leuchte passt in moderne aber auch ältere Wohngegenden mit Altbaubestand, sie fügt sich gut ein. Sehr schön die Möglichkeit der direkten und indirekten Ausleuchtung. Überzeugend die hohe Energieausbeute und die Reduzierung des Energiebedarfs um über 40 Prozent.
—

Judges panel
Considering all the regulations relating to vandal-proofing, modularity and suitability for retrofitting, the lamp is well designed. It is suitable for both modern residential districts and older areas with mature buildings; it fits in well. The fact that it can illuminate with direct and indirect light is very appealing. The high energy yield and the more than 40 per cent reduction in energy requirement are convincing arguments.
—

NightElements

Lichtsäulensystem
Lighting column system

Hersteller / Manufacturer
Hess Form + Licht GmbH
D-78050 Villingen-Schwenningen
—
Design / Designer
Vogt & Partner
CH-8400 Winterthur
—
Vertrieb / Distributor
Hess Form + Licht GmbH
D-78050 Villingen-Schwenningen
—

»NightElements« ist ein vielseitiges Säulensystem, das durch sein Baukastenprinzip alle gängigen Beleuchtungsanforderungen für Innenstädte erfüllt. Dazu gehören Straßen- und Platzbeleuchtung, verschiedene Anstrahlungen, Zonierungsbeleuchtung und Effekte.

Das System ist mit seinen rund 15 Lichtelementen weitgehend frei kombinierbar und lässt sich sehr flexibel gestalten. Dabei gibt es bei den verschiedenen Modulen auf unterschiedlichen Höhen Lichtaustritte nach oben, seitlich oder über 360°, symmetrisch oder asymmetrisch sowie mit schwenkbarem Spot. Bei den Zwischenmodulen ist die Länge je nach Bedarf frei wählbar.

Dieses Lichtsäulensystem ist eine Antwort auf die zunehmend kürzeren Nutzungsperioden im städtischen Außenraum. Die Möglichkeit, Leuchten zum Beispiel aus einem Park, der aufgegeben wird, später für eine Fassadenbeleuchtung zu gebrauchen, schont Ressourcen und Kosten. Das zeitlose Design ermöglicht spätere Erweiterungen oder Umnutzungen in jeder städtischen Umgebung, ohne störend zu wirken.

»NightElements« is a versatile lighting column system which, thanks to its modular principle, fulfils all the usual requirements for urban lighting, including lighting of streets and squares, various types of floodlighting, zone lighting and effects.

With some 15 lighting elements the system is to a large extent freely combinable and can be very flexibly arranged. The various modules have light outlets at different heights, upward, laterally or through 360 degrees, symmetrically or asymmetrically, and with swivelling spots. The length of the intermediate modules can be freely selected, according to requirements.

This lighting column system is a response to the ever-shorter periods of use in outdoor urban spaces. For example, using lamps from a disused park to light a façade economizes resources and cuts costs. The timeless design means that the system can later be extended or used for a different purpose in any urban environment without any disruptive effect.

Jury
Die Straßenlaterne folgt einem klaren und übersichtlichen modularen Entwurf. Die einzelnen Bauteile sind je nach Situation und Straßenverhältnissen flexibel einsetzbar. Man kann die Module in unterschiedlichen Höhen installieren, sodass sie verschiedene Funktionen erfüllen können. Eine sehr schlichte Lösung, die in ihren Proportionen klar auf das jeweilige Umfeld abgestimmt werden kann.
—

Judges panel
A street lamp with a clear, well-organized modular design. The components can be used in various ways, depending on the situation and street conditions. The modules can be installed at different heights to allow them to perform different functions. A very clear-cut solution, whose proportions can be clearly matched to the environment in question.
—

Grasshopper

Strahlerfamilie
Spotlight range

Hersteller / Manufacturer
Erco Leuchten GmbH
D-58507 Lüdenscheid
—
Design / Designer
Werksdesign / In-house design
Henk Kosche, Alois Dworschak
—
Vertrieb / Distributor
Erco Leuchten GmbH
D-58507 Lüdenscheid
—

Die neue Strahlerfamilie zeichnet sich durch kompakte, sehr effiziente Leuchtmittel und eine minimalistische, archetypische Anmutung aus. Sie verbindet Langlebigkeit in Design und Qualität mit der neuesten Technik der Lichterzeugung.

Mit Bestückungsvarianten in unterschiedlichen Wattagen sowie durch Reflektoren mit verschiedenen Abstrahlwinkeln lassen sich die »Grasshopper« Strahler auf individuelle Beleuchtungsaufgaben abstimmen oder im laufenden Betrieb anpassen. Neben Halogen-Metalldampflampen können Hochleistungs-LEDs in weiß oder in Varychrome-Technik für variable Lichtfarbe eingesetzt werden. Dabei kann farbig gestaltet werden und das mit nur geringem Energieverbrauch, der bisher durch den Einsatz von Farbfiltern sehr hoch war.

Mit robusten, mehrfach pulverbeschichteten Aluminiumgehäusen, dem fixierbaren Gelenk und vielseitigen Montagemöglichkeiten ist »Grasshopper« ideal, um Fassadendetails, Firmenschilder oder Skulpturen, Pflanzen oder Bäume akzentuiert zu beleuchten.

This new range of spotlights features compact, highly efficient lamps and a minimalist, archetypical appearance. It combines longevity in design and quality with leading-edge technology in light generation.

Lamp options with different wattages and reflectors with different beam angles allow the »Grasshopper« spotlights to be suited to individual lighting tasks or adjusted during operation. In addition to metal halide lamps, high-performance LEDs can be used, either in white or in Varychrome technology for variable colour of the light. This allows colour lighting with low energy consumption, which previously, due to the need for colour filters, was very high.

With robust, multiple powder-coated aluminium housings, the fixable joint and versatile mounting options, »Grasshopper« is ideal for spotlighting façade details, company nameplates, sculptures, plants or trees.

Jury
Die Leuchte wird für Outdoor-Beleuchtung in der Natur verwendet. Ungewöhnliches Lampendesign, sehr architektonisch. Das gibt einen schönen geometrischen Kontrast zur Weichheit der Natur. Eine gute Designlösung und eine gute Lösung für Energieeffizienz.
—
Judges panel
The light is used for outdoor lighting in nature. Unusual lamp design, very architectural. It results in a pleasant, geometric contrast to the softness of nature. A good design solution and a good solution for energy efficiency.
—

Lucendo
Solarleuchte / Solar-powered light

Lucendo

Solarbetriebene Hausnummernleuchte

Solar-powered house-number light

Hersteller / Manufacturer
Würth Solergy
D-71672 Marbach am Neckar
—
Design / Designer
Design Tech
Jürgen R. Schmid
D-72119 Ammerbuch
—
Vertrieb / Distributor
Würth Solergy
D-71672 Marbach am Neckar
—

»Lucendo« beleuchtet Hausnummern und -schilder unabhängig vom Stromnetz. Umweltschonende Photovoltaikmodule erzeugen aus Licht den notwendigen Strom und versorgen die Solarleuchte über einen Akku-Pack mit ausreichend Energie. Sie verbindet modernes funktionales Design mit bestechender Technologie. Statt der herkömmlichen Silizium-Module wurden innovative CIS-Module eingesetzt, die sich durch hohe Qualität und Lebensdauer sowie ein attraktives Erscheinungsbild auszeichnen.

Das reduzierte, auf die Ästhetik des Materials – mehrfach eloxiertes Aluminium – konzentrierte Design und die ausgewogenen Proportionen und Formen fügen sie harmonisch in verschiedenste architektonische Umgebungen ein.

»Lucendo Basic« erlaubt auch das Beleuchten von bereits bestehenden Hausnummern. »Lucendo Unikat« verbindet die Leuchte mit einer Acrylglasplatte, bei der die Beschriftung aus verschiedensten Farben und Schriftarten individuell gewählt wird. »Lucendo Special« schließlich wird für Sonderbeschriftungen, wie Firmenschilder, eingesetzt.

»Lucendo« illuminates house numbers and house-number plates independently of the grid. Eco-friendly photovoltaic modules generate the power needed from light and supply the solar light with sufficient energy via a rechargeable battery pack. The light combines modern, functional design with impressive technology. Instead of the conventional silicon modules, it uses innovative CIS modules, which stand apart by virtue of their high quality, long life and attractive appearance.

The reduced design, focusing on the aesthetics of the material—multiple-anodized aluminium—and the balanced proportions and forms allow it to fit harmoniously into a wide variety of architectural environments.

With »Lucendo Basic« existing house numbers can also be illuminated. »Lucendo Unikat« combines the light with an acrylic glass plate. The numbers and/or text on the plate can be chosen individually from a wide range of colours and fonts. »Lucendo Special« is a model suitable for special signs such as company nameplates.

Jury
Schöne Idee. Sehr energieeffizient, Solarenergie zu nutzen, um Hausnummern zu beleuchten. Für die Nachrüstung hervorragend, weil keine Verkabelung notwendig ist, sehr intelligente Lösung. Die Formensprache ist unprätentiös, langlebig und für die Beschilderung von Häusern geeignet.
—
Judges panel
Nice idea. Very energy-efficient to use solar power to light house numbers. Ideal for retrofitting because no wiring is needed—a very intelligent solution. The formal expression is unpretentious, long-lived and appropriate for signs on houses.
—

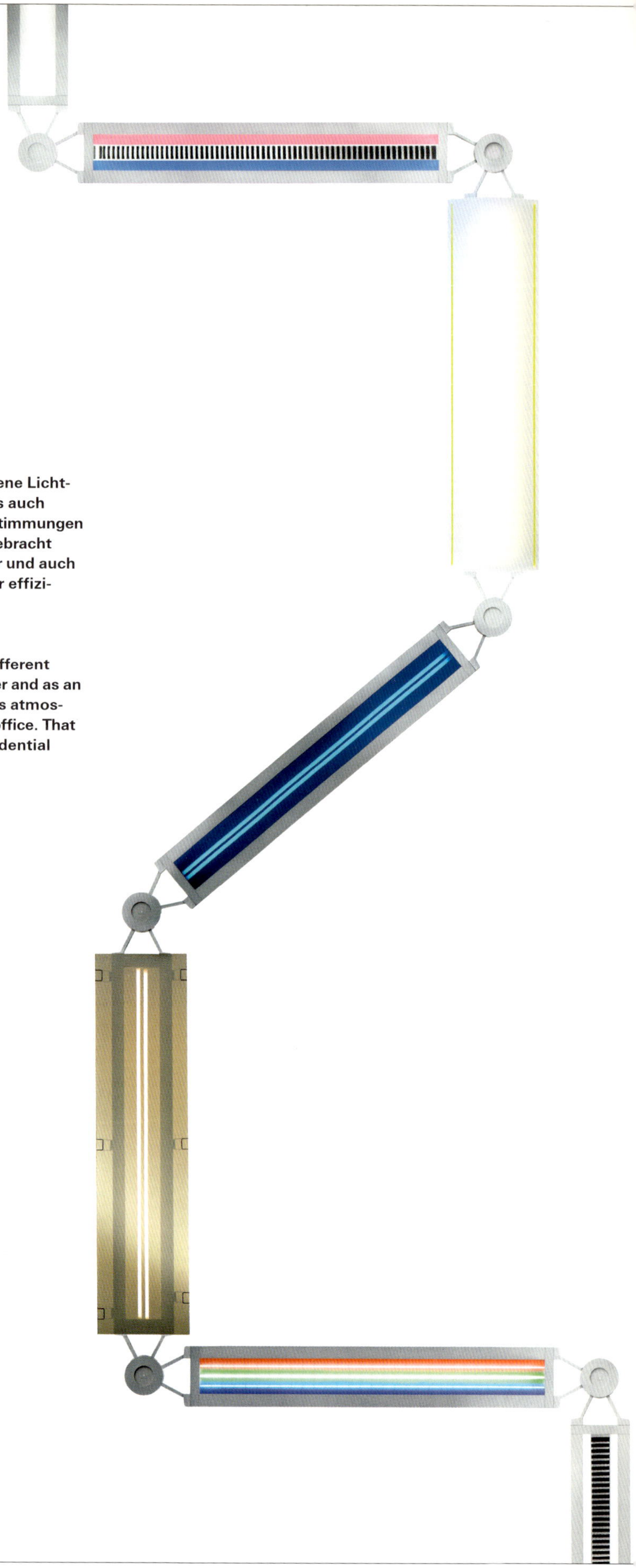

Jury

Variables, sehr flexibles System, das verschiedene Lichtatmosphären schafft, sowohl in der Reihung als auch als Einzelleuchte. Diese Büroleuchte lässt auch Stimmungen zu, die vielleicht nicht mit Büro in Verbindung gebracht werden. Sie wird dadurch vielseitiger einsetzbar und auch für den Wohnbereich nutzbar. Arbeitet mit sehr effizienten Leuchtstoffröhren.

—

Judges panel

Variable, highly versatile system that creates different lighting atmospheres, both when strung together and as an individual luminaire. This office light also allows atmospheres to arise which may be out of place in an office. That makes it more versatile, and also usable in residential areas. Uses very efficient fluorescent tubes.

—

BLUEmotion

Lineares Pendelleuchtensystem
Linear pendulum light system

Hersteller / Manufacturer
Trilux-Lenze GmbH + Co. KG
D-59759 Arnsberg
–
Design / Designer
Hartmut S. Engel
Design Studio
D-71634 Ludwigsburg
–
Vertrieb / Distributor
Trilux-Lenze GmbH + Co. KG
D-59759 Arnsberg
–

Das avantgardistische Leuchtensystem für T5 Hochleistungsleuchtstofflampen zeichnet sich durch die sinnliche Codierung der Lichtquellen und eine außergewöhnliche Bandbreite an lichttechnischem Repertoire aus.

So ermöglichen geregelte Lichtintensität und die Mischung von Lichttemperaturen, den natürlichen Lichtverlauf eines sonnigen Tages zu simulieren. »BLUEmotion« gestaltet Licht als emotionales Erlebnis. Sie bringt den Eindruck eines tiefblauen Sommerhimmels in die beleuchteten Räume. Das Raffinierte dabei ist, man erkennt blaue Lampen und erlebt weißes, sonniges Raumlicht. Dynamische Farblichtsteuerungen erschließen spezielle Lichtszenarien, die eine stimulierende und kreative Atmosphäre schaffen.

Auch die nahezu grenzenlose Modularität macht »BLUEmotion« einzigartig. Ein einziges Trägersystem und eine Vielzahl von optischen und ästhetischen Systemkomponenten bieten ein Maximum an gestalterischer Freiheit: Von der interessanten Einzelleuchte über nahtlose Lichtbänder bis hin zu geometrischen Figuren.

The distinctive features of this avant-garde luminaire system for T5 high-performance fluorescent lamps are the sensorial coding of the light sources and the exceptionally varied lighting technology repertoire.

For example, controlled luminous intensity and the mixing of light temperatures enable the natural course of light on a sunny day to be simulated. »BLUEmotion« presents light as an emotional experience, bringing the impression of a deep blue summer sky into the rooms it lights. What is ingenious is that one sees blue lamps, but experiences white, sunny interior lighting. Dynamic coloured light controls open up special lighting scenarios that give rise to a stimulating and creative atmosphere.

Its almost unlimited modularity also makes »BLUEmotion« unique. A single support system and a plethora of optical and aesthetic system components offer a maximum of design freedom – from an interesting individual luminaire through seamless lighting strips to geometrical forms.

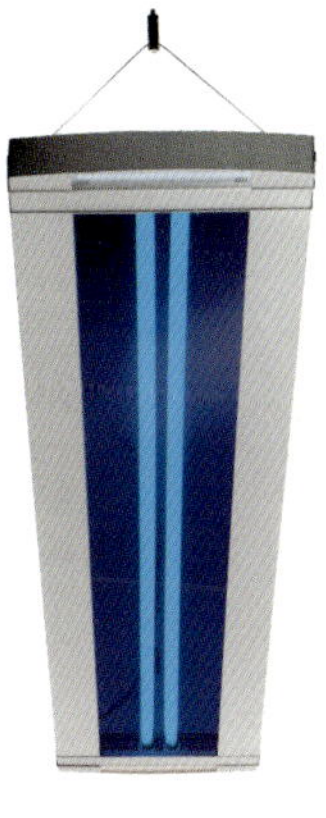

Jury

Sehr technisches Produkt, hochprofessionell, das sowohl in Altbauten wie in Neubauten einen wohltuenden Kontrast zur Architektur ermöglicht. Schafft eine sehr professionelle Arbeitslichtsituation. Besonders hervorzuheben ist die Kleinteiligkeit des Produkts, die durch ein Aneinanderreihen der kleinen Module trotzdem ermöglicht, beinahe endlose Lichtbänder zu schaffen.

—

Judges panel

Very technical product, highly professional, makes a welcome contrast to the architecture possible in both old and new buildings. Creates very professional lighting for work. Particularly commendable is the smallness: by stringing the little modules together it is nevertheless possible to create almost endless lighting strips.

—

Vaero-S

Pendelleuchtensystem
Pendulum light system

Hersteller / Manufacturer
Zumtobel Lighting GmbH
A-6850 Dornbirn
—
Design / Designer
Hartmut S. Engel
Design Studio
D-71634 Ludwigsburg
—
Vertrieb / Distributor
Zumtobel Staff Vertriebs GmbH
D-70567 Stuttgart
—

Das modulare Pendelleuchtensystem für energieeffiziente T5-Leuchtstofflampen garantiert erstklassige Arbeitsplatzbeleuchtung. Dafür sorgt eine revolutionäre Technologie: die Mikropyramiden-Strukturoptik MPO.

Basis sind feine und exakte Mikropyramiden, die für brillantes Licht und ausgewogene Kontrastverhältnisse sorgen sowie eine deutlich höhere Ausbeute des direkt abgestrahlten Lichts und reflexfreies Arbeiten an Bildschirm und Arbeitsplatz ermöglichen.

»Vaero-S« besticht nicht nur durch ihre zukunftsweisende Lichttechnik sondern auch durch klare Formensprache, optische Leichtigkeit und puristisches Design. Quadratisch strukturiert, ungewöhnlich flach und ein durchscheinendes Gehäuse sind Merkmale dieser herausragenden Büroleuchte.

Eine Innovation sind auch die in drei Positionen aufklappbaren Reflektorflügel. Sie steuern die Lichtwirkung im Raum und erlauben damit erstmals, Lichtverteilung und Leuchtdichte in Abstimmung auf die Aufhängehöhe anzupassen und ein optimales Kontrastverhältnis an der Decke zu erreichen.

A modular pendulum light system for energy-efficient T5 fluorescent lamps that guarantees first-class lighting of workplaces thanks to a revolutionary technology—micropyramid structure optics, or MPO.

The technology is based on fine, high-precision micropyramids that ensure brilliant light with balanced contrast, a significantly higher yield of directly radiated light, plus reflection-free working at monitors and desks.

The appeal of »Vaero-S« lies not only in the groundbreaking lighting technology, but also in its clear formal expression, apparent lightness and puristic design. A square structure, unusually flat, and a transparent housing are the characteristic features of this outstanding office light.

The reflector vanes, which can be unfolded in three positions, are also an innovation. They control the effect of the light in the room, making it possible for the first time to adapt the distribution and density of the light to the suspension height, and thus to achieve an optimal contrast ratio at the ceiling.

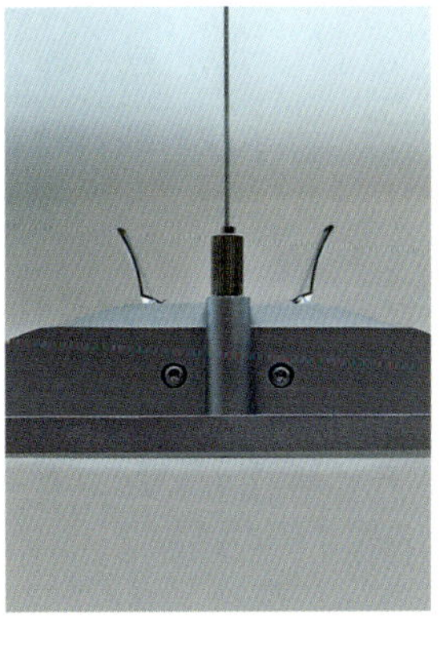

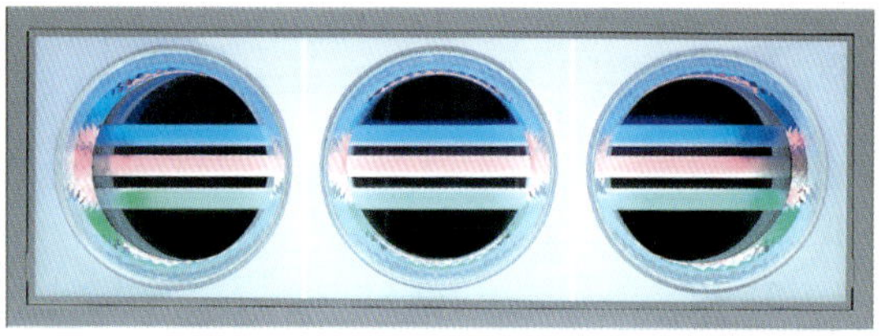

2Light
Downlight und Strahler / Downlight and spot

2Light

Downlight und Strahler
Downlight and spot

Hersteller / Manufacturer
Zumtobel Licht GmbH
D-32657 Lemgo
–
Design / Designer
Hartmut S. Engel
Design Studio
D-71634 Ludwigsburg
–
Vertrieb / Distributor
Zumtobel Staff Vertriebs GmbH
D-32657 Lemgo
–

»2Light« unterscheidet sich deutlich von herkömmlichen Downlightsystemen. Nicht allein wegen der großen Flexibilität des quadratischen Modulsystems, sondern vor allem wegen der einzigartigen, geradezu revolutionären Lichtwirkung.

Dabei verschmelzen der direkte und der indirekte Lichtanteil zu einer sanften, angenehm hellen und sinnlichen Lichtstimmung. Möglich wird dies durch das innovative Funktionsprinzip des so genannten Reflektorsplittings im Leuchtenkörper.

Einzigartig sind aber auch die Farblichtinszenierungen, die mit »2Light« möglich sind. So können zusammen mit intelligenter, digitaler Steuerungstechnik ohne größeren Aufwand dynamische, erlebnisorientierte Beleuchtungskonzepte umgesetzt und gezielt mit jedem beliebigen Ton des gesamten Farbspektrums Lichtstimmungen erzeugt werden.

Neuartig ist der staubdichte Leuchtenkörper, dessen Gehäuse komplett geschlossen ist. »2Light« gibt es in verschiedenen Varianten: als 3-, 2- und 1-Module sowie als 2Light Mini im Kompaktformat.

There is a world of difference between »2Light« and conventional downlight systems. Not just because of the great versatility of the quadratic modular system, but above all because of the unique, in fact revolutionary lighting effect.

The direct and indirect portions of the light merge to create a soft, pleasantly bright and sensuous mood. This is made possible by the innovative functional principle of »reflector splitting« in the lamp.

However, the coloured light scenarios that can be created with »2Light« are also unique. For example, with intelligent digital control technology it is not difficult to implement dynamic, experience-oriented lighting designs and to create specific lighting atmospheres with any desired shade chosen from the entire colour spectrum.

The dustproof lamp with its fully enclosed housing is also new. Different models of »2Light« are available – with 3, 2 or 1 modules and in a compact version, the »2Light Mini«.

Jury
Sehr gute, innovative Lösung, die Farbstimmungen ermöglicht. Überzeugend auch der Schutz vor Verstauben und Verschmutzen. Die Kombination der Verstellbarkeit der Strahler hinter dem Glas mit dem sehr reduzierten Rahmen sowie der Easy-Clean-Verschluss zum Öffnen des Glases ist gut gelöst.
–
Judges panel
Very good, innovative solution allowing colour moods. The protection against dust and dirt is also impressive. The combination of adjustability of the spots under the glass with the extremely reduced frame and the easy-clean lock for opening the glass is well designed.
–

Blade
LED Sicherheitsleuchte / LED safety light

Blade

LED Sicherheitsleuchte
LED safety light

Hersteller / Manufacturer
Sigma Elektro GmbH
D-67433 Neustadt
—
Design / Designer
Industrial PDD GmbH & Co. KG
D-70173 Stuttgart
—
Vertrieb / Distributor
Sigma Elektro GmbH
D-67433 Neustadt
—

Durch ihr modernes und flaches Design ohne scharfe Kanten ist »Blade« auch nach ihrem Einsatz beim Fahrradfahren sehr benutzerfreundlich; denn dann nimmt man diesen kompakten und formschönen Handschmeichler einfach in der Hosentasche mit und kann ihn anderweitig nutzen.

Möglich ist dies durch die flache Anordnung der Batterien, die in Analogie zur linearen LED-Reihung liegen. Drei konventionelle AA Batterien und drei weiße LEDs sorgen für eine extrem lange Betriebszeit und eine Leuchtkraft, die weit über die einer herkömmlichen Positionslampe hinausgehen. Durch die innovative Topfbauweise entsteht ein im Gesamtbild angepasstes Fugenbild, ohne dass die Lampe an Spritzwasserfestigkeit verliert.

»Blade« kann man mit nur einer Hand am Fahrrad anbringen und das in unterschiedlichen Positionen von horizontal bis vertikal, aufrecht oder kopfüber: denn die Verrastung an der Halterung ist in Stufen verstellbar. Die Halterung ist so konzipiert, dass die Lampe an jedem gebräuchlichen Lenker fixiert werden kann.

With its modern, slimline design, and with no sharp edges »Blade« is also very user-friendly after the cycle tour is over; because then you simply put this compact, well-designed light in your pocket and can use it for other purposes.

It is made possible by the flat arrangement of the batteries, analogous to the linear arrangement of the LEDs. Three conventional LR6 batteries and three white LEDs ensure an extremely long battery life and a luminosity that far exceeds that of a conventional position light. Thanks to the innovative enclosed construction the joints harmonize with the overall appearance without the lamp being less splashproof.

»Blade« can be attached to the bicycle with one hand and in different positions—from horizontal to vertical, upright or upside down, because the lock on the holder is adjustable in steps. The holder is designed to allow the lamp to be fixed to all ordinary handlebars.

Jury
Aufregende Innovation für Fahrräder. Guter Einsatz von energiesparenden LEDs. Eine sehr clevere Designlösung, da man die Leuchte am Fahrrad flexibel befestigen und auch einfach ins Büro oder nach Hause mitnehmen kann. So kann sie nicht gestohlen werden.
—
Judges panel
Exciting innovation for bicycles. Good use of energy-saving LEDs. A very clever design solution, because not only can you attach the light to the bike in various ways, you can also take it home or into the office. That way it can't be stolen.
—

LED UpDown Light

Einbauleuchte
Built-in light

Hersteller / Manufacturer
Paulmann Licht GmbH
D-31832 Springe
—
Design / Designer
Werksdesign / In-house design
Matthias Vogt
—
Vertrieb / Distributor
Paulmann Licht GmbH
D-31832 Springe
—

»UpDown Light« ist eine multifunktionale Mini-LED-Einbauleuchte, die dank modernster superflacher LED-Technik fast nicht eingebaut werden muss. In Verbindung mit einem speziellen Ring wird die LED ohne technischen Aufwand zu einer extrem flachen Leuchte, die vielseitig eingesetzt werden kann – auch nachträglich – beispielsweise in Parkett, Fliesen, Laminat oder Teppich aber auch in Möbel, im Feuchtraum sowie im geschützten Außenbereich.

Sehr feine Stromzuleitungen ermöglichen eine unsichtbare Verlegung der Kabel in kleinsten Zwischenräumen, wie Fliesenfugen oder Teppichkanten. Die Montage ist durch das Clicksystem einfach und selbsterklärend.

Die LEDs verbrauchen sehr wenig Strom, funktionieren wartungsfrei, haben eine lange Lebensdauer von 100 000 Stunden und schonen somit die Ressourcen. Durch die klare und schlichte Formensprache und durch viele Variationsmöglichkeiten passt die Leuchte in unterschiedlichste Umgebungen. Es gibt sie in satin, transparent, blau oder rot, Ring und Rahmen in Chrom matt oder Eisen gebürstet.

»UpDown Light« is a multifunctional mini-LED built-in light that hardly needs to be built in, thanks to the latest, extremely flat LEDs. In combination with a special ring, and without any additional technology, the LED is transformed into an extremely flat light that can be installed—or retrofitted—almost anywhere, for example in parquet, tiling, laminate or carpeting, as well as in furniture, bathrooms, and in covered outdoor areas.

The very thin power supply lines can be laid concealed in minute spaces such as tiling joints or along the edges of fitted carpets. Installation is simple and self-explanatory, thanks to the click system.

The LEDs consume very little power, are maintenance-free, have a long life of 100,000 hours, and thus help conserve natural resources. Thanks to its clear, uncluttered formal expression and the many possible variations, the light fits into a wide variety of different surroundings. It is available in satin, transparent, blue or red; the ring and frame in matt chrome or brushed iron.

Jury
Fußbodenbeleuchtung ist ein gängiger Standard in der Architektur. Sehr überzeugend ist, hier jetzt LEDs als Leuchtmittel einzusetzen. Das macht Sinn wegen der geringen Einbauhöhe, der Lichtstärke und auch, weil nicht nur weißes sondern auch farbiges Licht als Signalwirkung beispielsweise für Sicherheitszonen eingebaut werden kann.
—
Judges panel
Floor lighting is a common standard in architecture. It is a very good idea to use LEDs for lighting in this area. It makes sense in view of their flatness, their luminous intensity and because not only white but also coloured lights can be installed, as signals, e.g. for security zones.
—

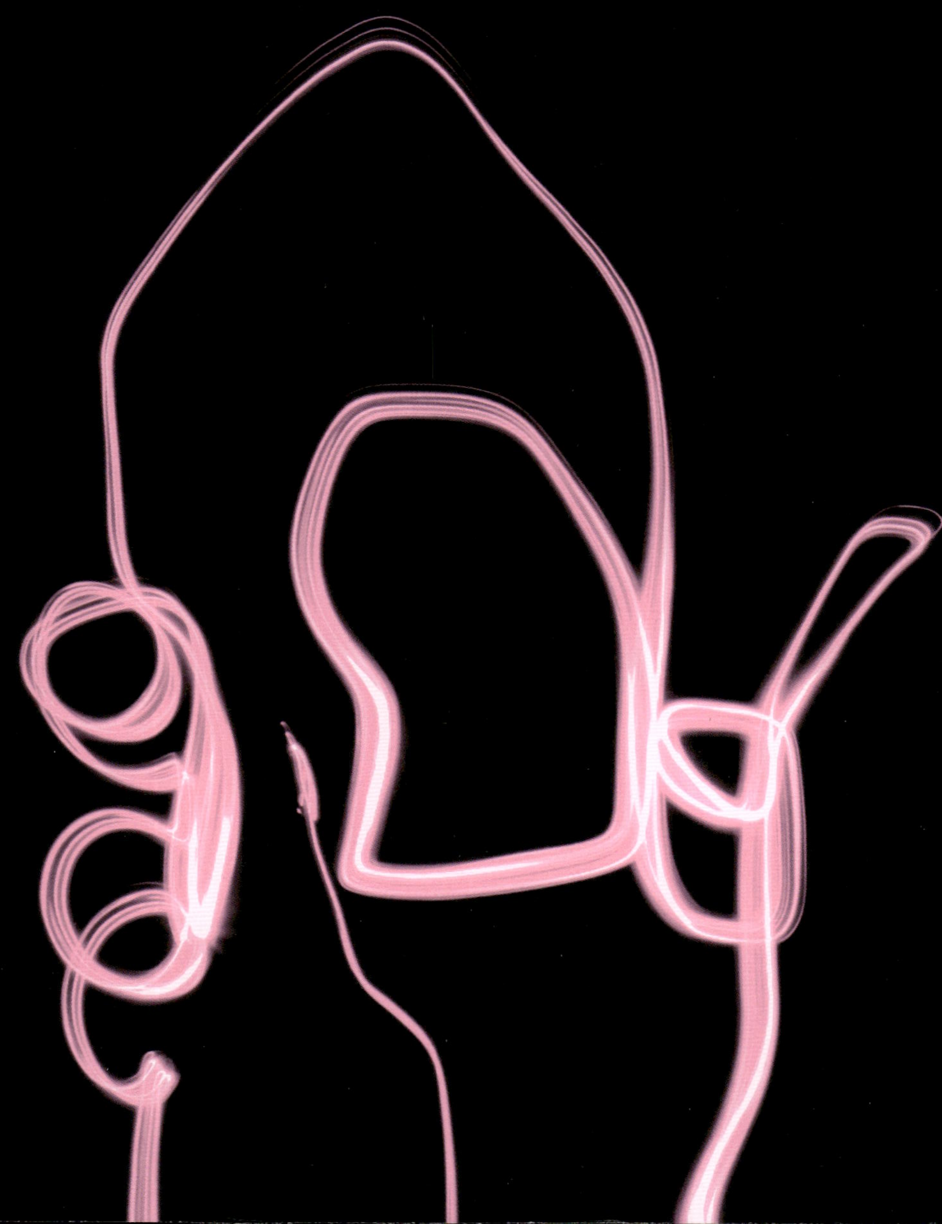

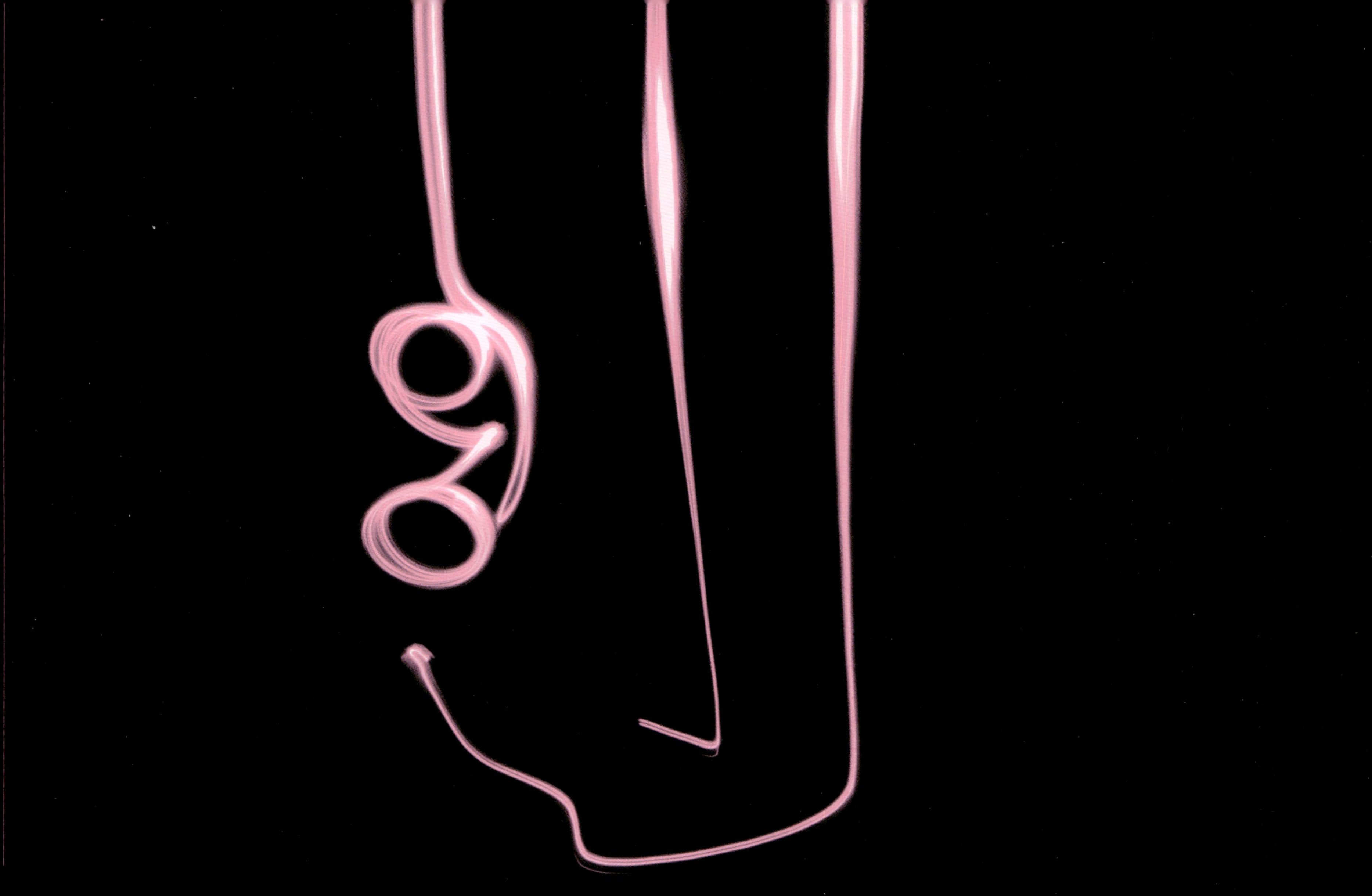

Panther
Flughafenlöschfahrzeug / Airport fire engine

Transport, Verkehr / Gold
Transport, traffic / Gold

Panther
Flughafenlöschfahrzeug / Airport fire engine

Panther

Flughafenlöschfahrzeug
Airport fire engine

Hersteller / Manufacturer
Rosenbauer International AG
A-4060 Leonding
—
Design / Designer
Spirit Design
A-1190 Wien
—
Vertrieb / Distributor
Rosenbauer Deutschland
D-76185 Karlsruhe
—

Das Design des »Panther« spiegelt nicht nur seine Verlässlichkeit und Sicherheit, die das Fahrzeug bei der Brandbekämpfung bietet, sondern auch seine Dynamik, Kraft und Stärke. Das Cockpit verfügt über eine Panorama-Windschutzscheibe aus Verbundsicherheitsglas und transparente Türen. Die weit nach oben und unten gezogenen Front- und Seitenscheiben bieten dem Fahrer einen optimalen Überblick über seine Umgebung. Elektrische Außenschwingtüren schließen automatisch bei einer Geschwindigkeit von 5 km/h. Dank eines optimierten Entwicklungsprozesses und einer intelligenten Gestaltung der Karosseriebauteile konnten in der Produktion Zeit und Ressourcen gespart werden.

Der 12 Meter lange, 3 Meter breite und 3,6 Meter hohe größere »Panther MA5 8x8« verfügt über einen 1000 PS Motor mit Allison-Automatik-Getriebe und ist damit ein energiegeladenes Produkt. Das Fahrzeug beschleunigt mit einem Betriebsgewicht von 40 000 Kilogramm in 25 Sekunden vom Stand auf 80 Stundenkilometer und erreicht eine Höchstgeschwindigkeit von 145 Stundenkilometern.

Der kleinere »Panther MA5« hat die Maße 11,8 x 3,0 x 3,5 Meter und verfügt über eine Motorleistung von 740 PS. Mit einem Betriebsgewicht von 31 000 Kilogramm beschleunigt er in 31 Sekunden von Null auf 80 Stundenkilometer. Seine Höchstgeschwindigkeit beträgt 120 Stundenkilometer.

The design of the »Panther« reflects not only the reliability and safety of this vehicle when fighting fires, but also its dynamics, power and strength. The cab features a panorama windscreen of laminated safety glass and transparent doors. The windscreen and side windows extend high and low, offering the driver an excellent view of the surroundings. Electrically powered swing-out doors close automatically when the vehicle's speed exeeds 5 km/h. An optimized development process and intelligent design of the body components made time savings and lower use of resources possible in production.

The larger model, the »Panther MA5 8x8«, is 12 metres long, 3 metres wide and 3.6 metres high. With a 1,000 bhp engine and an Allison automatic transmission it is a veritable powerhouse. Despite an operating weight of 40,000 kg, the fire engine accelerates from 0 to 80 km/h in 25 seconds and has a top speed of 145 km/h.

The smaller model, the »Panther MA5«, measures 11.8 x 3.0 x 3.5 metres and has a 740 bhp engine. It has a service weight of 31,000 kg and takes 31 seconds to accelerate from 0 to 80 km/h. Its top speed is 120 km/h.

Jury
Faszinierende Formgebung, da sie auf Feuerwehrleute zugeschnitten ist. Es ist nicht wirklich mit rationalen Argumenten zu erklären, aber es zeigt, welche Formgestaltungsmerkmale soziokulturell möglicherweise prägend sind. Ein sehr energiereicher Entwurf, der auch für groben Einsatz die richtige Formensprache findet. Gut ausgeführtes, funktionales Design. Eine lebensrettende Maschine.
—

Judges panel
Fascinating design, because tailored to the needs of firefighters. It is not really explicable with rational arguments, but it shows which design features may make an impression socioculturally. A very energetic design, with the right formal expression, even for tough assignments. Well executed, functional design. A life-saving machine.
—

Jury

Überzeugendes Bionik-Konzept. Es beginnt mit dem Kofferfisch. Das Prinzip dieser bionischen Struktur wird auf ein Fahrzeug übertragen. Dadurch wird der c_W-Wert deutlich reduziert. Beeindruckend die Anwendung der Form des Kofferfisches, um den Energieverbrauch zu reduzieren.

—

Judges panel

An impressive study in bionics. It starts with the cow-fish. The principle of this bionic structure is transferred to a vehicle. As a result, the drag coefficient is substantially reduced. The use of the shape of the cow-fish to reduce energy consumption is impressive.

—

Mercedes-Benz Bionic Car

Konzeptfahrzeug
Design study

Hersteller / Manufacturer
DaimlerChrysler AG
D-70546 Stuttgart
–
Design / Designer
DaimlerChrysler Design Team
D-71063 Sindelfingen
–
Vertrieb / Distributor
DaimlerChrysler
Vertriebsorganisation Berlin
D-10878 Berlin
–

Die Natur erreicht ihre Ziele stets ökonomisch und mit einem Minimum an Energie. Vorbild für das Konzeptfahrzeug ist der Kofferfisch, der trotz seines kantigen, würfelförmigen Rumpfes hervorragende Strömungseigenschaften hat und deshalb ein aerodynamisches Ideal darstellt.

Mit einem c_W-Wert von 0,19 zählt es zu den strömungsgünstigsten Automobilen dieser Größenklasse. Dies trägt zusammen mit zukunftsweisender Dieselmotorentechnik und einer neuartigen Abgasreinigung maßgeblich zur Kraftstoffeinsparung von bis zu 20 Prozent und zur Verringerung der Stickoxid-Emissionen von bis zu 80 Prozent bei.

Der Kofferfisch ist auch ein Musterbeispiel für Steifigkeit und Leichtbau. Seine Außenhaut besteht aus sechseckigen Knochenplatten, die so gewachsen sind, dass sie bei minimalem Gewicht ein Höchstmaß an Festigkeit bieten und ihn wirksam vor Verletzungen schützen. Das wurde auf das Konzeptfahrzeug übertragen. So lässt sich beispielsweise an der Außenhaut der Türen eine bis zu 40 Prozent höhere Steifigkeit erzielen. Berechnet man die gesamte Rohbaustruktur nach dem bionischen Muster, sinkt das Gewicht bei unverändert guter Stabilität und Crashsicherheit um rund ein Drittel.

Nature always achieves her aims economically and with a minimum of energy. The model for this design study is the cow-fish, which despite its angular, dice-shaped rump has excellent flow characteristics and therefore represents an aerodynamic ideal.

With a drag coefficient of 0.19 it is one of the most streamlined cars in its class. Together with pioneering diesel technology and an innovative exhaust gas treatment system, this adds up to a fuel saving of up to 20 per cent and a reduction of up to 80 per cent in NOx emissions.

The cow-fish is also a classic example of rigidity and lightweight construction. Its outer skin consists of hexagonal bone plates, which grow in such a way that they have maximum strength with minimum weight, and protect the fish effectively against injury. All this was transferred to the design study. Thus, for example, the outer skins of the doors can be made up to 40 per cent more rigid. If the entire bodyshell is calculated in accordance with the bionic pattern, the weight is reduced by about a third, with the same good stability and crash safety.

Symbiotische Optik

LED Frontscheinwerfer für Kraftfahrzeuge

LED front headlight for motor vehicles

Hersteller / Manufacturer
Schefenacker AG
D-71409 Schwaikheim
—
Design / Designer
Werksdesign / In-house design
Alf Liesener
—
Vertrieb / Distributor
Schefenacker AG
D-71409 Schwaikheim
—

LEDs sind effizient in der Energieausnutzung, wartungsfrei und leuchten ein Autoleben lang. Der Einsatz verschiedener Module in einem Scheinwerfer erlaubt eine exakt definierte Ausleuchtung der Fahrbahn, und durch ihre markanten Optiken eröffnen LED-Scheinwerfer völlig neue Designperspektiven.

Zwar ist LED-Technik bereits als Tagfahrlicht in Frontscheinwerfern auf dem Markt, für Abblend- und Fernlicht besteht aber noch Entwicklungsbedarf. Gerade hier hat die Firma eine Reihe optischer Konzepte entwickelt, die das LED-Licht besonders effizient nutzen, zum Beispiel die »symbiotische Optik«, die zu mehr Lichtstärke führt. Hier gelingt es, die magische 50 Prozent-Hürde bei der optischen Effizienz des Abblendlichts zu überschreiten.

Bei der Gestaltung des Prototyps stand die Natur Pate. Wie Blütenblätter gruppieren sich Blendenringe um die glitzernden und technisch anmutenden LED-Optiken. Dies unterscheidet sich gezielt von herkömmlichen Scheinwerferlösungen, die sich in der Regel geometrischer Grundformen bedienen. Die im Prototyp eingesetzten Optiken zeichnen sich sowohl durch sehr hohe Effizienz als auch durch hohe Präzision der Bauteile aus.

LEDs are energy-efficient, maintenance-free and shine for a car's entire lifetime. The use of different modules in a headlight makes precisely defined illumination of the carriageway possible, and LED headlights with their striking optics open up completely new dimensions in design.

LED technology is already on the market in daytime running lamps in headlights. For dipped and high beam lights, however, further development is needed. It is in this area that the company has developed a number of optical concepts that use LED light particularly efficiently, e.g. the »symbiotic optics«, which result in higher light intensity. With this technology the magic 50 per cent mark in the optical efficiency of dipped beam light has been exceeded.

The design of the prototype was inspired by nature. Diaphragm rings are grouped like petals around the glittering, very technical-looking LED optics. This distinguishes them deliberately from conventional headlights, which usually make use of basic geometric shapes. The optics used in the prototype stand out due to their very high efficiency and the high precision of the components.

Jury
Obwohl die LED-Technologie natürlich schon existiert, wurde hier ein innovatives, auch auf dem Vorbild der Natur basierendes System angewendet. Sehr interessant die Innovation und das Engineering. Zukunftsweisendes Produkt. Das Zukunftsdesign wird das Autodesign beeinflussen, aber auch die Energieeffizienz: der Scheinwerfer verbraucht weniger Energie und ist wartungsfrei.
—

Judges panel
LED technology already exists, of course, but here an innovative system has been used which is also based on examples found in nature. Both the innovation and the engineering are very interesting. Ground-breaking product. The futuristic design will influence not only car design but also energy efficiency, because the headlight consumes less energy and is maintenance-free.
—

Sun Set
Solarbetriebener Stromversorger / Solar power unit

Sun Set

Mobiler solarbetriebener Stromversorger

Mobile solar power unit

Hersteller / Manufacturer
Würth Solergy
D-71672 Marbach am Neckar
—
Design / Designer
KPG Design Group GmbH
D-73527 Täferrot
—
Vertrieb / Distributor
Würth Solergy
D-71672 Marbach am Neckar
—

»Sun Set« ist eine mobile, solarbetriebene Stromversorgungseinheit. Sie dient zur Versorgung elektrischer Kleinverbraucher in einer Umgebung ohne Netzstrom. Die Energiequelle ist dabei Sonnenlicht. Konzipiert wurde es für den Einsatz in Dritte-Welt-Ländern zur Beleuchtung von Hütten, aber auch in Katastrophengebieten oder in Orten ohne Stromversorgung: wie für die Beleuchtung einer Gartenlaube oder eines Carports, das Betreiben einer Gewächshausbelüftung, sowie im Freizeit-/Outdoorbereich. Angewendet werden kann es für Alarmsysteme, tragbare Funk-, Fernseh- und Radiogeräte, für Kühleinheiten, Lüftungsanlagen und medizinische Notversorgungsgeräte.

Über ein Würth Solar CIS-Modul wird die integrierte Akkumulatoreneinheit geladen. Von dort werden dann Kleingeräte (1,5 bis 12 V) mit elektrischer Energie tageszeit- und witterungsunabhängig versorgt. Alle Prozesse, von der Umwandlung des Sonnenlichts in Energie bis hin zur Abgabe an den Verbraucher, werden elektronisch gesteuert. Das führt zu einem optimierten Wirkungsgrad, zu Langlebigkeit und Bedienerfreundlichkeit.

»Sun Set« besitzt zudem nicht nur zwei Ladeplätze für Energiesparleuchten, die jederzeit abgenommen werden können, sondern auch zwei weitere Ladeplätze zum Aufladen von handelsüblichen Kleinbatterien (Akkus) sowie eine 12 Volt-Steckdose.

The »Sun Set« mobile solar power unit provides power for small electrical appliances in areas where there is no power grid. The energy source is sunlight. »Sun Set« was originally intended for hut lighting in Third World countries, as well as for use in disaster areas or places with no power supply, for lighting arbours or carports, for powering greenhouse ventilation systems, and in general for use in leisure/outdoor activities. It can be used for alarm systems, portable radios and TVs, for refrigeration and ventilation units, and for emergency medical care devices.

The power unit's rechargeable battery is charged by a Würth Solar CIS module. Independently of the time of day and weather conditions, the battery in turn supplies electricity—1.5 to 12 volts—to power small appliances. All processes are electronically controlled, from the conversion of the sunlight into energy to delivery of the power to the consumer units. The result is optimized efficiency, long life and user-friendliness.

In addition, »Sun Set« not only has two charging stations for energy-saving lanterns, which can be removed at any time, but also two charging bays for standard rechargeable batteries and a 12-volt power outlet.

Jury
Es wurde schon vor über 10 Jahren versucht, Solarenergie im Campingbereich einzuführen. Das hat nicht geklappt, weil die Produkte rein auf funktionaler Ebene gestaltet worden sind. Die Ikonographie der alten Coleman-Gasleuchten hier dagegen aufzugreifen, erscheint der Jury als ein guter Versuch, Solartechnologie im Outdoor-Bereich anzuwenden. Überzeugend, auch die Funktion des Bügels.
—

Judges panel
An attempt was made more than 10 years ago to introduce solar energy in the camping sector. It didn't work, because the products were designed purely on a functional level. By contrast, taking up the iconography of the old Coleman gas lamps seems a good attempt to apply solar technology in the outdoor sector. Makes good sense—the way the handle works, as well.
—

Freizeit, Outdoor / Silber
Leisure, outdoor activities / Silver

Waterbelt
Flaschenhalter / Bottle holder

Waterbelt

Flaschenhalter
Bottle holder

Hersteller / Manufacturer
DNS Designteam
D-70178 Stuttgart
—
Design / Designer
Werksdesign / In-house design
Ralph Schäflein + Kai-Uwe Neth
—
Vertrieb / Distributor
DNS Designteam
D-70178 Stuttgart
—

»Ohne Wasser kein Leben, keine Lebensenergie« – »Waterbelt« ermöglicht, bei jeder Gelegenheit eine (Wasser-) Flasche bei sich zu führen. Gerade heute, wo die Menschen immer mobiler sind, haben viele eine Flasche Wasser bei sich. Dafür bietet der neuartige praktische Flaschenhalter die ideale Lösung. Man braucht nicht zusätzlich eine Tasche mitzuführen und hat die Hände frei.

»Waterbelt« besteht aus einem längenverstellbaren Gurt und einem Flaschenclip mit »Easy Clip Off«-Funktion. Der Clip – in schwarz oder weiß glänzend – passt für fast alle handelsüblichen PET- und Glasflaschen mit einem Halsdurchmesser bis circa 31 Millimeter und lässt sich durch leichten Druck auf die Griffmulde einfach lösen.

Der Gurt ist wahlweise aus Polyester oder Nylon, kann in allen Farben hergestellt und nach Wunsch bedruckt werden.

Jury
Ohne Wasser keine Energie. Ein simples und leicht verständliches Alltagsprodukt, das es uns möglich macht, ohne unnötiges Beiwerk die PETflasche mit dem Trinkwasservorrat für den ganzen Tag umgehängt stets bei sich zu führen. Auch als nettes Give Away auf Messen, Kongressen oder als Werbeartikel bestens geeignet.
—
Judges pane
No water, no energy. A simple, easily understandable everyday product without any unnecessary embellishments that enables us to take our PET bottle with the day's drinking water supply with us wherever we go. Also suitable as a nice give-away at trade fairs, conferences or as promotional gifts.
—

»No water, no life, and no vital energy« — »Waterbelt« makes it possible to always carry a (water) bottle. Especially nowadays, in this age of ever-increasing mobility, a lot of people carry a bottle of water. And for that purpose the new, practical bottle holder is the ideal solution. It isn't necessary to carry an extra bag, and both hands are free.

»Waterbelt« consists of a belt of adjustable length and a bottle clip with an »easy clip-off« function. The clip — in black or glossy white — fits almost all commonly available PET and glass bottles with neck diameters of up to approx. 31 millimetres, and is easy to release by applying light pressure to the recess in the handle.

The belt is made of polyester or nylon to choice, can be produced in any colour and printed as desired.

Jury

Sehr überzeugende und innovative Idee, zum Antrieb des Unterwasserfahrzeugs Druckluft aus dem Tauchsport zu nehmen. Guter Einsatz von Technologie und Material, guter Einsatz von Energie mittels Druckluft. Kompaktes dynamisches Design.

—

Judges panel

Very impressive and innovative idea to use compressed air—as used by divers—to propel this underwater vehicle. Good application of technology and materials, good application of energy by means of compressed air. Compact, dynamic design.

—

Twin Diver 3000

Unterwasserscooter
Underwater scooter

Hersteller / Manufacturer
Twin Diver AG
D-89231 Neu-Ulm
–
Design / Designer
Eisele Kuberg Design
D-89231 Neu-Ulm
–
Vertrieb / Distributor
Pauli's Tauchshop
D-89250 Senden
–

Herkömmliche Tauchscooter brauchen Strom und müssen immer wieder länger an die elektrische Ladestation. Sie bewegen sich ohne Stopp – entweder auf- oder abwärts.

Der »Twin Diver 3000« dagegen bekommt seine Energie von Druckluft, die Tauchern zur Verfügung steht. Er ist mit einem weltweit neuartigen Druckluftantrieb ausgestattet. Zentrale Innovation ist der hoch entwickelte, pneumatische Radialkolbenmotor. Er bringt volle Leistung, ist umweltfreundlich und wartungsarm.

Mit dem »Twin Diver 3000« kann man sinken, steigen, anhalten, wie man möchte, und er hat eine extrem kurze Auftankdauer. Der Unterwasserscooter mit einer Maximalgeschwindigkeit von 5 Stundenkilometern kann vielseitig eingesetzt werden: sowohl für Hobbytaucher als auch als Rettungs- und Arbeitsmobil für Profitaucher. Außerdem ist er mit einem Sicherheits-Geschwindigkeitsventil mit Auto-Stopp-Funktion bei Kontaktverlust ausgestattet.

Der seewasserresistente, kompakte Unterwasserscooter besitzt ergonomisch günstige Griffzonen. Angeformte Haltegriffe und vertiefte Armaturen verringern die Verletzungsgefahr.

Conventional underwater scooters are powered by electricity and have to dock regularly at a charging station. They move—either upward or downward—without stopping.

By contrast, the »Twin Diver 3000« is powered by compressed air, which divers have at their disposal. It has a compressed air propulsion system—the first of its kind in the world. The central innovation is the advanced pneumatic radial piston motor. It delivers full performance, is environmentally friendly and needs little maintenance.

The »Twin Diver 3000« allows its users to descend, rise and stop at will, and can be refuelled very quickly. With a top speed of 5 km/h, the underwater scooter is an all-rounder, suitable both for amateur divers and as a rescue and working vessel for professionals. It also has a safety speed valve with an autostop function if users lose contact with it.

The compact, sea water-resistant underwater scooter has ergonomically well-placed handling zones. Moulded handles and recessed controls reduce the risk of injury.

Jury

Kleines Spaßgerät zum Schwimmen und Tauchen. Dynamische Gestaltung. Leicht transportierbar. Wird mit einem Elektromotor betrieben, der offensichtlich keine Verschleißteile hat. Auf Grund des Elektromotors hat das Gerät keine Emissionen.

–

Judges panel

A little funmobile for swimming and diving. Dynamic design. Easy to transport. Is propelled by an electric motor that evidently has no wear-and-tear parts. Because the motor is electric the apparatus is emission-free.

–

Seabob

Wassersportgerät
Aquatic sports apparatus

Hersteller / Manufacturer
Rotinor GmbH
D-70499 Stuttgart
—
Design / Designer
Schweizer Design Consulting
D-70182 Stuttgart
—
Vertrieb / Distributor
Rotinor
D-33609 Bielefeld
—

Mit dem »Seabob« kann man rasant durch das Wasser fahren, entspannt gleiten oder tauchen. Sporttaucher können damit bis 40 Meter tief tauchen, in der Serienprogrammierung ist dies aus Sicherheitsgründen auf 2,5 Meter begrenzt.

Das Sportgerät mit Jet-Antrieb erreicht über Wasser eine Geschwindigkeit bis zu 20 Stundenkilometern. Gesteuert wird der »Seabob« einfach durch Gewichtsverlagerung des Körpers. Angetrieben wird er durch ein weltweit patentiertes Elektro-Jetstream-System. Dank der einzigartigen Elektrotechnologie ist das Wasserfahrzeug leistungsfähiger als andere Systeme, zugleich kostengünstig im Betrieb und umweltfreundlich. Der Motor arbeitet frei von Emissionen, ist nahezu lautlos und wartungsfrei.

Die Energie liefern qualitativ hochwertige High-Energy-Li-Ion-Akkumulatoren. Die zukunftsweisende Li-Ion-Technology stammt aus der Weltraumforschung. Der »Seabob« verbindet damit modernste High-Technology mit einem ansprechenden Design. Dank seiner dynamischen Formgebung ist er im Wasser außerordentlich beweglich.

The »Seabob« allows its user to speed through the water, to relax and glide along, or to dive. Sports divers can dive to depths of 40 metres with it. In the standard program the diving depth is limited to 2.5 metres for safety reasons.

On the surface, this aquatic sports apparatus with jet propulsion attains speeds of up to 20 km/h. To steer the »Seabob«, users simply shift the weight of their bodies. It is propelled by an electro-jetstream system, patented worldwide. Thanks to this unique electrical technology the vessel performs better than other systems, while at the same time being relatively cheap to operate and eco-friendly. The motor is emission-free, virtually silent and maintenance-free.

The energy is supplied by high-quality, high-energy rechargeable lithium-ion batteries. The pioneering lithium-ion technology is a result of space research. Thus the »Seabob« combines leading-edge advanced technology with an attractive, dynamic design which gives it exceptional manoeuvrability in the water.

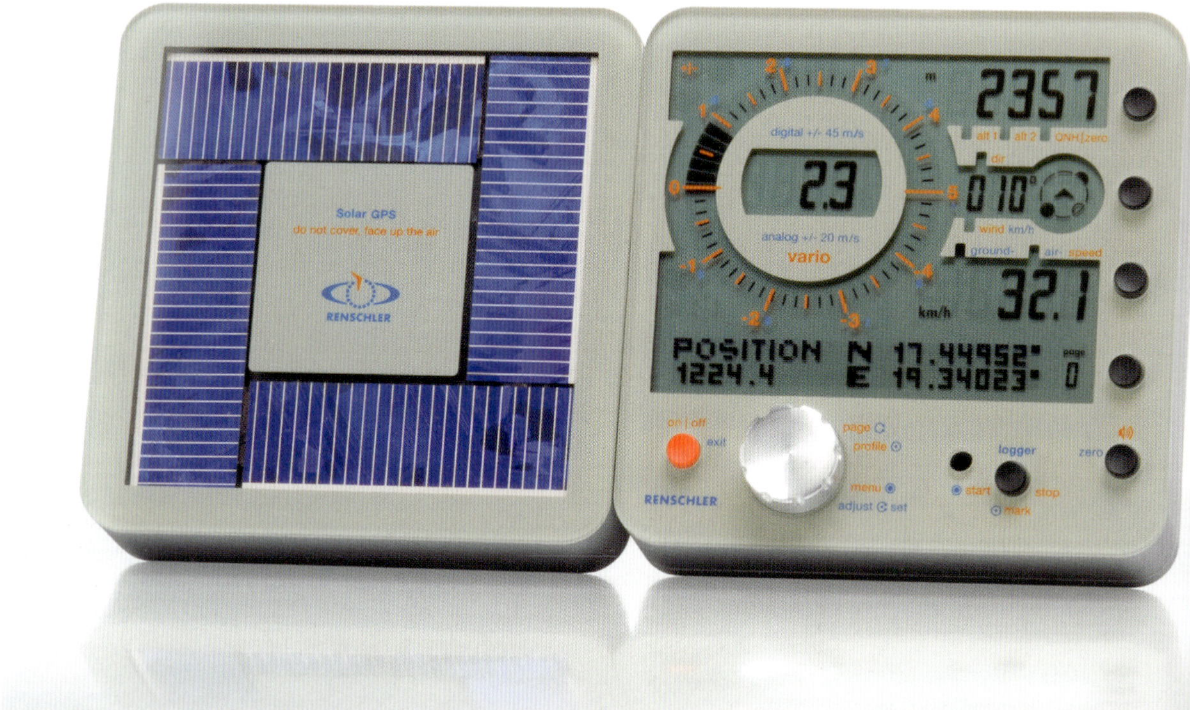
Solar GPS
do not cover, face up the air
RENSCHLER
digital +/- 45 m/s
2.3
analog +/- 20 m/s
vario
m
2357
alt 1
alt 2
QNH | zero
dir
010°
wind km/h
ground-
air- speed
km/h
32.1
POSITION N 17.44952°
1224.4 E 19.34023°
page
0
on | off
exit
page
profile
menu
adjust set
logger
start
stop
mark
zero
RENSCHLER

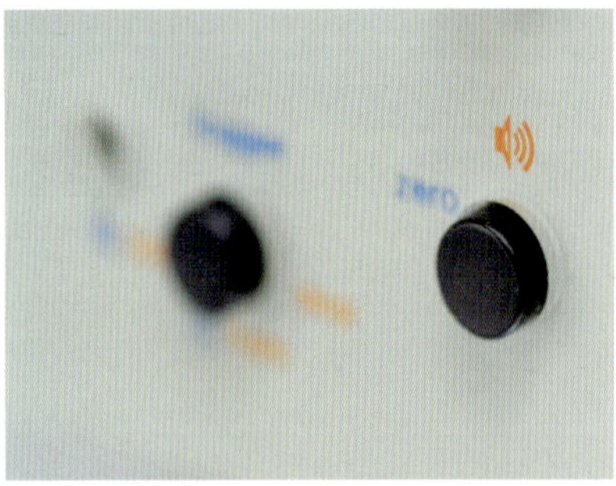

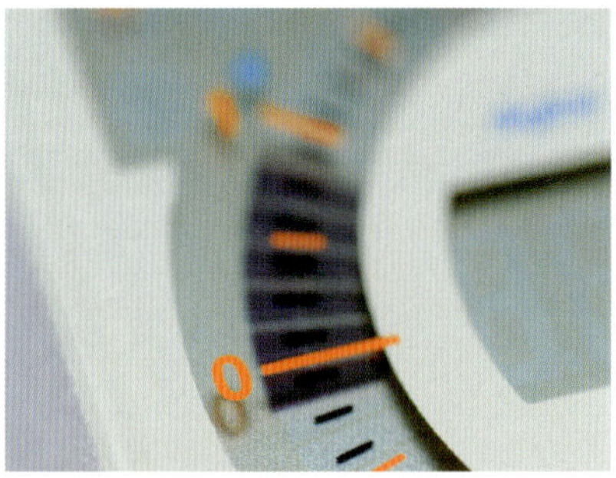

010°
wind km/h
ground-

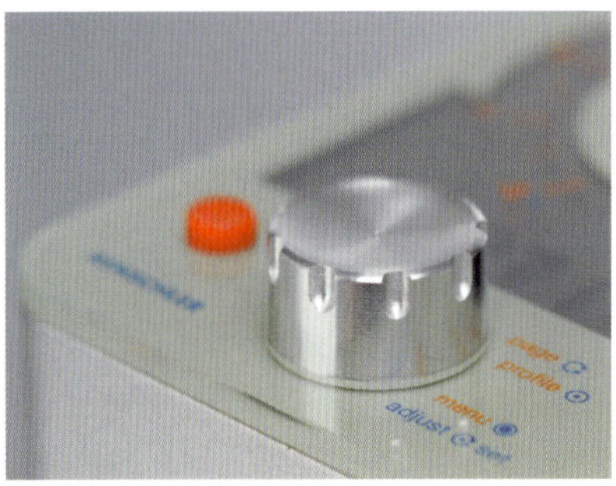

CoMo

Fluginstrument
Flight instrument

Hersteller / Manufacturer
Renschler
D-70435 Stuttgart
—
Design / Designer
Jacob Bachmayer
D-70191 Stuttgart
—
Vertrieb / Distributor
Renschler
D-70435 Stuttgart
—

Cockpit-Modul, kurz »CoMo«, heißt die neue, modular erweiterbare Fluginstrumentenserie mit einem innovativen Konzept. Die Cockpit-Module und Add-on-Module können nach eigenem Ermessen miteinander kombiniert werden.

Die solare Stromversorgung macht das »CoMo« zu einem der besten und sparsamsten Fluginstrumente auf dem Markt für Drachen- und Gleitschirmflieger, für Ballonfahrer und Segelflugpiloten, insbesondere in Kombination mit einem der Add-on GPS-Module. Es ist besonders benutzerfreundlich. So ist zum Beispiel jeder Funktionsanzeige auf dem Display eine eigene Taste zugeordnet. Das erleichtert die Handhabung und schließt Fehlbedienungen weitgehend aus.

Eine Neuheit sind die frei programmierbaren Flugprofile, womit man zum Beispiel für Thermikflug, Talquerung oder Acroflug eine spezielle Geräte-Einstellung abrufen kann. Das elegante und robuste Aluminiumgehäuse, die klar gegliederte, kontrastreiche Anzeige, sowie die leicht zu bedienende, handschuhtaugliche Tastatur mit dem Rotary-Einstellknopf zeichnen das Design von »CoMo« aus.

Cockpit-Modul, or »CoMo« for short is the name of a new, modular flight instrument series with a novel design concept. The cockpit modules and add-on modules can be combined at the user's discretion.

Its solar power supply makes the »CoMo« one of the best and most economical flight instruments on the market for the pilots of hang gliders and paragliders, of hot-air balloons and gliders, in particular in combination with one of the add-on GPS modules. It is exceptionally user-friendly. For example, a dedicated key is assigned to each function shown on the display. This facilitates handling and virtually rules out mistakes in operating the instrument.

The freely programmable flight profiles are completely new. They allow special settings to be called up for, e.g., thermal soaring, valley overflights or special settings. Distinctive design features of »CoMo« include the elegant and robust aluminium housing, the clearly arranged, contrast-rich display and the keypad with rotary setting knob, easy to operate even with gloves on.

Jury
Große Weiterentwicklung verglichen mit dem Stand der Technik und des Design. Weil man verschiedene Module hat, die aneinandergehängt und gut montiert werden können. Insofern wird diese ganze Unordnung, die ein Paragleiter normalerweise vor sich hat, aufgeräumt. Gestalterisch gut gelöst, betrieben mit Solarenergie. Diese Innovation überzeugt.
—
Judges panel
Major further development, compared with the state of the art in technology and design, because there are different modules that can be coupled together and mounted well. Thus the clutter that paragliders normally have in front of them is tidied up. Well designed, solar-powered. An impressive innovation.
—

Jury

Interessante Interpretation des Eternitpflanzencontainers. Durch unterschiedliche Möglichkeiten der Aufstellung ergeben sich ganz verschiedene Bilder in der Außenpräsentation von Pflanzen. Recyclebar. Überzeugt.

—

Judges panel

Interesting interpretation of the Eternit plant container. The various positioning options result in outdoor plant arrangements with completely different appearances. Recyclable. Convincing.

—

Twista

Pflanzengefäß
Plant container

Hersteller / Manufacturer
Eternit AG
CH-8867 Niederurnen
–
Design / Designer
Martin Mostböck
A-1010 Wien
–
Vertrieb / Distributor
Schneider & Held GmbH
D-88481 Balzheim
–

»Twista«, das innovative Gestaltungselement für Gartenarchitekten, Landschaftsplaner und den kreativen Anwender, sticht durch seine ungewöhnliche Form hervor, die sich stark von seiner Umgebung abhebt und Dynamik vermittelt.

Der Entwurf basiert auf speziell gekrümmten Flächen, die einen Körper mit ausgeprägten architektonischen und plastischen Qualitäten bilden. Eine bewegte Form entsteht, die entweder als Einzelobjekt aufgestellt oder mit mehreren zu einem System zusammengefügt werden kann.

Dabei bieten sich vielfältige Kombinationsmöglichkeiten, die sich formal im Dualismus von Kante und Welle abbilden. Ähnlich der Bewegung eines Grashalmes im Wind bringt »Twista« Spannung in die Pflanzen- und Gartenlandschaft. Die Pflanzengefäße, die es in zementgrau oder anthrazit gibt, sind aus Eternit, das zu 100 Prozent recyclebar ist, hergestellt.

»Twista«, the innovative design unit for landscape gardeners, landscape planners and creative users, stands out due to its unusual form, which contrasts strongly with its surroundings and conveys a dynamic impression.

The design is based on specially curved surfaces that form a corpus with pronounced architectural and three-dimensional qualities. The result is a dynamic form that can either be stood individually or grouped to create a system.

In the latter case there is a wide variety of possible combinations, which find formal expression in the dualism of edge and wave. Rather like the movement of a blade of grass in the wind, »Twista« introduces a tension into plant and garden landscapes. The plant containers, available in cement grey or anthracite, are made of Eternit, which is 100 per cent recyclable.

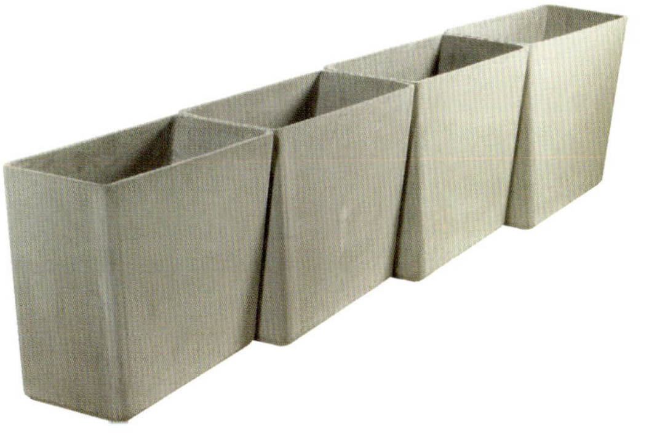

Westerland

Taschen
Bags

Hersteller / Manufacturer
BREE Collection GmbH & Co. KG
D-30916 Isernhagen
—
Design / Designer
BREE Design Team
Tim Knossalla
—
Vertrieb / Distributor
BREE Collection GmbH & Co. KG
D-30916 Isernhagen
—

Bei »Westerland« stand der Ostfriesennerz Pate. So sind die Taschen denn auch aus dem robusten Material der leuchtenden Regenjacken: synthetisches Material mit Teflonbeschichtung und textilem Trägermaterial. Das beschichtete Cotton-Gewebe ist besonders griffig und angenehm.

Die Tasche ist schlauchartig gearbeitet. Der obere Teil wird abgeklappt, mit Klipp-Verschlüssen gesichert und schützt den Tascheninhalt so wirkungsvoll vor Wind, Wetter und Langfingern. Ein breiter, verstellbarer Gurt sorgt neben dem anschmiegsamen Material für hohen Tragekomfort. Im Innern überzeugt die übersichtliche funktionale Gestaltung.

»Westerland« gibt es in zwei Farben – Sun/Marine und Jeans/Marine – und sechs verschiedenen Modellen: drei Schultertaschen, zwei Messengerbags und einer Einkaufstasche, die sogar wendbar ist.

»Westerland« was inspired by those well-known rain-jackets known as »East Friesian mink«. The bags also are made of the same robust, bright-coloured synthetic material, with a Teflon coating and textile backing. The coated cotton fabric is particularly pleasant to the touch.

The bag is fashioned like a tube. The top section is folded over, secured with clip fasteners, and thus protects the contents of the bag effectively against wind, weather and pickpockets. Besides the comfortable material, a broad adjustable strap helps ensure high wear comfort. The well-arranged, functional interior is also a convincing argument.

»Westerland« comes in two colours—sun/marine and denim/marine—and six different models: three shoulderbags, two messenger bags and a shopper, which is even reversible.

Jury
Interessante Interpretation des alten Ostfriesennerzes. Aus dem gleichen gelben Material. Die Einkaufstasche ist wie das Vorbild von beiden Seiten zu benutzen. Die Taschen verkörpern sehr schön die Anmutung, wenn man im Regen an der Küste entlangläuft. Von der Funktionalität her gut gelöst.
—
Judges panel
Interesting interpretation of the old »East Friesian mink«. Made of the same yellow material. And, like the original, the shopper can be used both sides. The bags very nicely convey the feeling of walking along the coast in the rain. Well thought out as regards functionality.
—

Jury

Junge Interpretation des Trolleys, mal nicht nur in Schwarz, sondern auch in verschiedenen Farben. Hat eine Anmutung wie eine Sporttasche. Junge Zielgruppe für klassischen Trolley wird angesprochen. Sehr funktional durch unterschiedliche Fächer, zum Beispiel ein Fach für einen Laptop. Die Stange kann weggepackt werden, dann lässt sich der Trolley als Tasche oder als Koffer tragen.

—

Judges panel

A young interpretation of the trolley, in various colours besides black. Has something of a sports bag about it. Addresses a young target group for the classic trolley. Highly functional, thanks to various compartments, for example a compartment for a laptop. The handlebar can be stowed away, and then the trolley can be carried like a bag or suitcase.

—

Punch

Reise- und Freizeittaschen
Travel and leisure bags

Hersteller / Manufacturer
BREE Collection GmbH & Co. KG
D-30916 Isernhagen
–
Design / Designer
BREE Design Team
–
Vertrieb / Distributor
BREE Collection GmbH & Co. KG
D-30916 Isernhagen
–

»Punch« ist mittlerweile ein Klassiker. Seit 10 Jahren ist die erfolgreiche, preisgekrönte Taschenserie auf dem Markt. Aus diesem Anlass wurde sie nicht nur überarbeitet, sondern auch um neue Modelle in den Bereichen »Reise« und »Sport« erweitert. Neben einem Trolley wurden entsprechende Taschen herausgebracht.

Die Taschen bestehen aus einem industriellen, extrem strapazierfähigen, fast unverwüstlichen Planenmaterial, das von einem Gewebeträger gestützt wird. Das hochwertige und modern verarbeitete Material wurde jetzt eine Spur dünner und damit noch leichter.

Das Tascheninnere zeichnet sich durch seine übersichtliche und detailreiche Aufteilung aus, die immer für Ordnung sorgt. »Punch« gibt es in den Farben Milk, Dark Orange und Black.

»Punch« is meanwhile a classic. This successful, award-winning range of bags has been on the market for 10 years. To mark the anniversary it has not only been given a facelift but also expanded with new models in the »travel« and »sports« categories. Besides a trolley, matching bags a travel and a sports bag have been launched.

The bags are made of an extremely hard-wearing, almost indestructible industrial tarpaulin supported by a fabric backing. This high-quality material, processed by modern methods, has now been made a tad thinner and thus even lighter.

The interior of the bag excels with very clear and detailed partitioning that ensures tidiness. »Punch« is available in three colours: milk, dark orange and black.

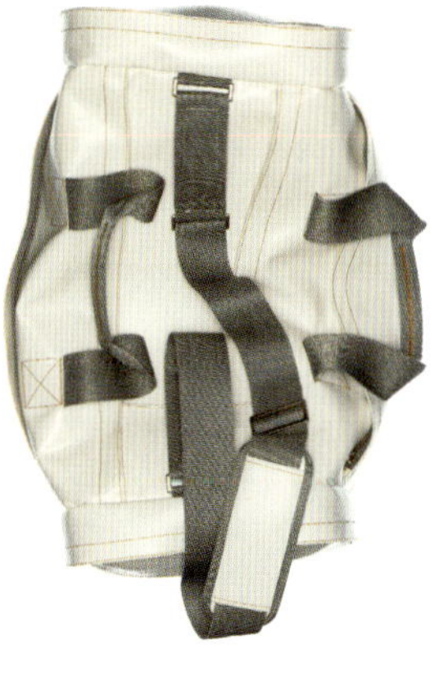

GAMELLE
GAMELLE
GAMELLE

Gamelle

Taschen aus Original-Gamellen der Schweizer Armee

Bags made from original Swiss Army mess tins

Hersteller / Manufacturer
Design Atelier Maurer GmbH
CH-9055 Bühler
–
Design / Designer
Werksdesign / In-house design
Walter Maurer
–
Vertrieb / Distributor
Handelsagentur CDH
D-72074 Tübingen
–

Das Einzelkochgeschirr, wie die Gamelle offiziell heißt, führte die Schweizer Armee bereits 1898 ein. Seine Form ist seitdem unverändert geblieben. Dieses Essgeschirr, das heute noch jeder Schweizer Soldat beim Eintritt in die Armee bekommt, besitzt einen eisernen Tragbügel, an dem man es über das Feuer hängen kann. Am Handgriff des Pfannendeckels ist eine Schlaufe angebracht, mit der man die Gamelle mittels Lederriemen festbinden kann.

Aus den gebrauchten, in der Armee nicht mehr verwendeten Gamellen wurden die intakten Modelle umgerüstet und mit modischen Farben einbrennlackiert. Entstanden sind modische Taschen, die zu Kultobjekten werden können.

Aus folgenden Farben kann ausgewählt werden: schwarz, weiss, orange, hellgrün, rosa, rot, dunkelgrün, dunkelblau. Außerdem gibt es noch eine rote Variante mit einem Schweizerkreuz. Mit einem verstellbaren Lederträger wird die Gamelle umgehängt. Innen wurde sie mit einem blauen Filzfutter ausgekleidet, in das eine Innentasche integriert ist. Im Deckel wurde ein kleiner Spiegel eingearbeitet.

The Swiss Army introduced the individual cooking utensil, as the mess tin is officially known, as early as 1898. Its shape has remained the same ever since. This mess tin, still issued today to every Swiss soldier joining the ranks, has a steel hanger by means of which it can be hung over the fire. A loop on the handle of the pan lid allows the mess tin to be fastened with leather straps.

The intact mess tins no longer used in the army have been converted and stove-enamelled in fashion colours. In this way they have been transformed into fashionable bags that could become cult objects.

There is a choice of the following colours: black, white, orange, light green, pink, red, dark green and dark blue. In addition there is a red version with a Swiss cross. The mess tin is slung over the wearer's shoulder with an adjustable leather strap. The interior is lined with a blue felt lining with an inner pocket. There is a small mirror mounted in the lid.

Jury
Sehr überzeugendes Recycling-Konzept, alte Armeebehälter zu nehmen, mit einem neuen Anstrich zu überziehen und sie dann als Taschen für ein jüngeres Publikum zu verkaufen. Durchaus sehr intelligent, ökologisch orientiert. Und die Funktionalität ist natürlich völlig in Ordnung, weil es in langjährigem Armeegebrauch war. Insgesamt ein sehr lustiges, interessantes Objekt, das Kultstatus erreichen könnte.
–

Judges panel
A very impressive recycling idea to take old army canisters, give them a fresh coat of paint and sell them as bags for a younger consumer target group. Thoroughly intelligent and ecologically oriented. And the functionality is fine, of course, because the product was in use in the army for years. All in all a very amusing, interesting object that could achieve cult status.
–

Ed Annink
Produkt- und Grafikdesigner,
Miteigentümer und Creative Director von
Ontwerpwerk Den Haag, Niederlande
–
Ed Annink
Product designer and graphic designer,
co-owner and creative director of Ontwerpwerk,
The Hague, Netherlands
–

Nach seiner Ausbildung an der Gold- und Silberschmiedeschule in Schoonhoven studierte Ed Annink Möbeldesign und Innenarchitektur an der königlichen Kunst- und Designakademie in Den Haag. Neben seiner Arbeit als Produkt-, Ausstellungs- und Grafikdesigner hat er sich vor allem einen Namen als Dozent gemacht. Er unterrichtete an verschiedenen Design- und Kunsthochschulen, unter anderem an der Hochschule der Künste Berlin, der Design Academy in Eindhoven und der Kunstakademie Arnheim, dem Istituto Europeo di Design in Mailand und der Université de Montreal in Quebec.

In der Fachwelt wird Ed Annink besonders wegen seines freien, experimentellen Ansatzes und seines interdisziplinären Designverständnisses geschätzt, das er mit 30 Mitarbeitern in seinem eigenen Studio Ontwerpwerk in Den Haag in die gestalterische Praxis umsetzt. Das Studio arbeitet für so renommierte internationale Unternehmen wie Driade, Authentics, Frankfurter Messe, Droog Design, Designum, Royal VKB, Nexus Technology, ExperimentaDesign, Purple South, Showroom.

After finishing his studies at the School for Goldsmiths, Silversmiths and Jewellers in Schoonhoven, Ed Annink studied furniture design and interior architecture at the Royal Academy of Art in the Hague. In addition to his work as a product, exhibition and graphic designer, he has also made a name for himself as a lecturer. He has taught at a number of design schools and art colleges including Berlin University of the Arts, the Design Academy in Eindhoven and the Arnheim Institute of the Arts, Istituto Europeo di Design in Milan and Université de Montreal in Quebec.

In the design world, Ed Annink is particularly well known for his free experimental approach and interdisciplinary design work, which he produces in cooperation with 30 employees in his own studio, Ontwerpwerk, in the Hague. The studio's clients include famous international companies such as Driade, Authentics, Messe Frankfurt, Droog Design, Designum, Royal VKB, Nexus Technology, ExperimentaDesign, Purple South, and Showroom.

Christoph Böninger
Industrie-Designer, Gesellschafter der Agentur brains4design, München
—
Christoph Böninger
Industrial designer, partner in brains4design, a design agency in Munich
—

Christoph Böninger studierte Industrie-Design in München und Los Angeles und arbeitete von 1983 bis 1987 in dem renommierten Designbüro Schlagheck & Schultes Design in München. Seine berufliche Laufbahn führte ihn anschließend in die USA, wo er bis 1990 den Bereich Design bei Siemens USA leitete. Danach verantwortete er die Produktentwicklung von Investitionsgütern der Siemens AG von München aus. Von 1997 bis 2006 war er Geschäftsführer der designafairs GmbH in München – einem der größten Studios in Deutschland. Seit Juli 2006 ist er Gesellschafter der Agentur brains4design in München.

Nebenbei nimmt er verschiedene Lehraufträge an Hochschulen wahr und ist seit 1996 im Vorstand im Industrie Forum Design Hannover. Seit 2004 ist er sowohl Vorstandsvorsitzender des Design Zentrum München als auch im Kuratorium der Zollverein School of Management and Design Essen. Produkte, an deren Gestaltung Christoph Böninger beteiligt war, sind vielfach mit internationalen Designpreisen ausgezeichnet worden, so auch im Design Center Stuttgart mit einem Focus in Silber im Jahr 2005.

Christoph Böninger studied industrial design in Munich and Los Angeles and worked from 1983 to 1987 in the reputed design office of Schlagheck & Schultes Design in Munich. His career path then took him to the USA where he led the Siemens US design team. After this spell abroad he came back to manage the product development of capital goods at Siemens AG in Munich. From 1997 to 2006 he was the general manager of designafairs GmbH in Munich—one of the largest design studios in Germany. He has been a partner in the brains4design agency in Munich since July 2006.

In addition, he lectures at a number of design colleges and has been on the management board of Industrie Forum Design in Hanover since 1996. Since 2004 he has been both the chairman of the management board of the Design Zentrum in Munich and on the Zollverein School of Management and Design in Essen. Products which Christoph Böninger has been involved in have received awards at numerous international design competitions, including the Design Center Stuttgart, where he received a Focus in Silver in the year 2005.

Henning Horn
Grafikdesigner,
Leiter des Design Center Stuttgart
–
Henning Horn
Graphic designer,
Head of Design Center Stuttgart
–

Nach einer Ausbildung zum Industriekaufmann in der Druckindustrie studierte Henning Horn Grafikdesign. Als Diplom-Designer arbeitete er in Designbüros in Kopenhagen und London, bei der Audi AG in Ingolstadt und in einer Frankfurter Corporate Identity Beratung. Danach machte er sich mit seinem Büro Formkontakt Corporate Design Management in Stuttgart selbständig. Sein Schwerpunkt ist die Integration verschiedener Designdisziplinen in umfassende Gestaltungsprozesse.

Seit 1998 leitet er das Design Center Stuttgart. Dort ist er unter anderem für das Konzept der jährlich wechselnden thematischen Orientierung bei den Focus-Wettbewerben und für die Planung und Ausrichtung der Internationalen Grafikdesign-Konferenzen »Face-to-Face« verantwortlich. Außerdem ist er Initiator der bundesweit einheitlichen Standards für den Zertifikatslehrgang »Design-Management (IHK)«, Fachjuror in zahlreichen Innovations- und Design-Wettbewerben im In- und Ausland und war bis 2006 gewählter Sprecher der Konferenz der Deutschen Designzentren.

After a business training for a career in the printing industry, Henning Horn studied graphic design. As a designer he worked in design studios in Copenhagen and London, with Audi AG in Ingolstadt and for a Frankfurt consulting firm specializing in corporate identity. After that he went freelance with his own studio, Formkontakt Corporate Design Management in Stuttgart. His main interest is the integration of different design disciplines in complex design processes.

He has been head of the Design Center Stuttgart since 1998, where he is responsible, among other things, for the concept of the Focus competitions with their annually changing themes, and for planning and organizing »Face to Face«, the international graphic design conference. He has also initiated nationally uniform standards for Chamber of Commerce courses in Design Management, has been a judge in many national and international innovation and design competitions, and was elected spokesman of the conference of German design centres until 2006.

Dr. phil. Rita Pohle,
Industrial und Interior Designerin, Autorin und Feng-Shui-Expertin,
design & consulting, Stuttgart.
–
Dr. phil. Rita Pohle,
Industrial and interior designer, author and Feng Shui expert,
design & consulting, Stuttgart.
–

Rita Pohle studierte zunächst Germanistik und Politologie an der Freien Universität Berlin und schloss ihr Studium mit dem Magister ab. Anschließend absolvierte sie ein Industrial Design-Studium an der Hochschule der Künste in Berlin. Nach ihrem Abschluss als Industrial Designerin HdK arbeitete sie dort als Wissenschaftliche Mitarbeiterin im Fachgebiet Ergonomie. Von 1992 bis 1996 arbeitete sie als Referentin im Design Center Stuttgart und promovierte mit dem Thema »Ästhetisches Handeln im Schrebergarten« zum Dr. phil. an der Freien Universität Berlin. Von 1997 bis 1999 lehrte sie als Gastprofessorin an der Fachhochschule Hannover im Fachbereich Design und Medien.

Ihr eigenes Büro, design & consulting, betreibt sie seit 1996 in Stuttgart. Ihre Schwerpunkte sind dabei die Gestaltung von Arztpraxen, Geschäfts- und Privaträumen nach Feng Shui, Farbkonzepte sowie Beratung und Coaching. Sie ist nicht nur Autorin zahlreicher Bücher zum Thema Feng Shui, sondern hält auch viele Seminare und Vorträge.

Rita Pohle initially studied German and politics at the Freie Universität Berlin, and graduated with a master's degree. Thereafter she completed her studies in industrial design at the Hochschule der Künste in Berlin. After graduating from the HdK as an industrial designer she began working there as an assistant in ergonomics. From 1992 to 1996 she worked as a consultant for the Design Center Stuttgart and was awarded her doctorate from the Freie Universität Berlin for her paper on »Aesthetic Approaches to the Garden Plot«. From 1997 to 1999 she was a guest professor at the Faculty of Design and Media at the University of Applied Sciences in Hanover.

She has run her own design office, design & consulting, in Stuttgart since 1996. Her design focus is on the design of doctors' surgeries, offices and private homes in line with Feng Shui principles and colour concepts. She also provides consulting services and coaching. Not only has she written numerous books on the topic of Feng Shui, she also holds a number of seminars and talks on the subject.

Frederick Rickmann
Geschäftsführer von Designnord,
Aarhus, Dänemark

—

Frederick Rickmann
General manager of Designnord,
Aarhus, Denmark

—

Frederick Rickmann gehört zu den Pionieren, die in den späten achtziger und frühen neunziger Jahren den Einsatz von IT Technologie und 3D CAD in Skandinavien vorangetrieben haben. Er ist heute Geschäftsführer von Designnord, einem etablierten führenden Designstudio, das von Aarhus und Taiwan aus für Kunden in aller Welt tätig ist. Designnord gestaltet im Bereich Produktdesign elektronische und medizinische Geräte und ist für Unternehmen in der Kommunikationsbranche tätig. Darüber hinaus hat das Studio zahlreiche Industrieanlagen entwickelt und Nutzfahrzeuge wie einen Traktor und einen Fischkutter. Es ist außerdem Spezialist für Interfacelösungen und bedient damit die Telekommunikation und andere Unternehmen, die in der Informationstechnologie tätig sind. Der Erfolg des Studios spiegelt sich in zahlreichen Designpreisen wider.

Frederick Rickmann engagiert sich auch intensiv im dänischen Designerverband, dessen Präsident er bereits gewesen ist, und in der Design Innovation Management Association (DIMA).

Frederick Rickmann is one of the pioneers and main drivers behind the use of computer technology and 3D CAD in Scandinavia in the late eighties and early nineties. Currently he is the general manager of Designnord, a leading design studio which works for clients from all over the globe from its bases in Aarhus and Taiwan. In the field of product design, Designnord designs electronic and medical devices and its clients include companies from the communications sector. Above and beyond this, the studio has developed countless industrial machines and commercial vehicles such as a tractor and a fishing trawler. Moreover, it is a specialist for interface solutions, serving the telecommunications sector and other companies in the field of information technology. The success of the studio is reflected in the numerous design prizes it has won.

Frederick Rickmann is also intensely involved in Danish Designers, an association where he has already served as president, and in the Design Innovation Management Association (DIMA).

Ass. Prof. Ursula Tischner
Architektin und Industrie-Designerin, econcept, Agentur für nachhaltiges Design, Köln. Associated Professor an der Design Academy Eindhoven, Niederlande
–
Ass. Prof. Ursula Tischner
Architect and industrial designer, econcept, agency for sustainable design, Cologne. Associated Professor at the Design Academy Eindhoven, Netherlands
–

Die Architektin und Industrie-Designerin arbeitete nach ihrem Studium, bei dem sie sich auf nachhaltige Gestaltung von Produkten und Dienstleistungen – Sustainable Design – spezialisierte, vier Jahre am Wuppertal Institut für Klima, Umwelt und Energie. 1996 gründete sie in Köln econcept, Agentur für nachhaltiges Design. econcept berät Unternehmen, führt Forschungsprojekte durch, entwickelt und gestaltet Produkte und Dienstleistungen und ist in der Aus- und Weiterbildung tätig. Das Ziel ist dabei immer, die ökonomische, ökologische und soziale Nachhaltigkeit von Produktions- und Konsumsystemen zu vergrößern.

Außerdem ist Ursula Tischner als Expertin für Sustainable Design in verschiedenen Netzwerken, Jurys und Gremien, wie zum Beispiel ISO oder DIN, tätig. Gemeinsam mit ihr hat das Design Center Stuttgart bereits erfolgreiche Projekte umgesetzt, zum Beispiel den internationalen Design-Kongress »Towards sustainable development« im Jahr 2000 und die Veranstaltung »ecobiente« 2005. Seit 2002 lehrt sie an der Design Academy Eindhoven. Außerdem ist sie Autorin zahlreicher Veröffentlichungen.

After her studies, during which she specialized in the design of sustainable products and services, the architect and industrial designer Ursula Tischner worked for four years at the Wuppertal Institute of Climate, Environment and Energy. In 1996 she founded econcept, agency for sustainable design, Cologne. econcept provides consulting services to companies, conducts research projects, develops and designs products and services and also provides training. The aim of the agency is to expand the economic, ecological and social sustainability of production and consumer systems.

Moreover, as an expert for sustainable design, Ursula Tischner is involved in various networks, juries and committees, such as the ISO and DIN standards boards. In cooperation with Ursula Tischner, the Design Center Stuttgart has completed a number of projects, such as the international design congress »Towards sustainable development« in the year 2000 and the »ecobiente« event in 2005. She has been teaching at the Design Academy in Eindhoven since 2002. Moreover, she is the author of numerous publications.

Was mehr als einem nützt Mia Seeger Preis 2006

Benefiting more than the individual Mia Seeger Prize 2006

Für junge Designerinnen und Designer – Absolventen der Studiengänge Industriedesign / Produktgestaltung, Innenarchitektur / Möbeldesign, Investitionsgüter- oder Transportation Design an deutschen Hochschulen:

Mia Seeger Preis 2006

Auszeichnung für Diplom- oder vergleichbare Entwurfsarbeiten. Voraussetzung: Diplom zwischen 2004 und 2006

Preise: 10.000 €

Jury
Prof. Karin Kirsch, Stuttgart
Heike Schnabel,
Schnabel, Schneider Industrial Design, Schorndorf
Michael Daubner,
Burkhardt Leitner constructiv GmbH, Stuttgart
Tom Schönherr,
Phoenix Design, Stuttgart
Stefan S. Handt,
DaimlerChrysler AG, Sindelfingen

Einsendeschluß: 12. Juni 2006

Ausführliche Wettbewerbsunterlagen:
Mia Seeger Stiftung
c/o Design Center Stuttgart
Willi-Bleicher-Straße 19
70174 Stuttgart
T +49 7 11 1 23 27 81
miaseegerstiftung@web.de

Im Focus Energy:
Mia Seeger Preis gemeinsam mit dem Internationalen Designpreis Baden-Württemberg

For young designers—graduates from courses in Industrial Design/Product Design, Interior Design/Furniture Design, Capital Goods Design or Transportation Design at German universities:

Mia Seeger Prize 2006

Award for undergraduate dissertations or comparable design work. Condition: Contestants must have graduated between 2004 and 2006

Awards: 10,000 €

Judges panel
Prof. Karin Kirsch, Stuttgart
Heike Schnabel,
Schnabel, Schneider Industrial Design, Schorndorf
Michael Daubner,
Burkhardt Leitner constructiv GmbH, Stuttgart
Tom Schönherr,
Phoenix Design, Stuttgart
Stefan S. Handt,
DaimlerChrysler AG, Sindelfingen

Deadline for submission of entries: 12 June 2006

For more details of the competition contact:
Mia Seeger Stiftung
c/o Design Center Stuttgart
Willi-Bleicher-Straße 19
70174 Stuttgart
T +49 7 11 1 23 27 81
miaseegerstiftung@web.de

In Focus Energy:
Mia Seeger Prize together with the Baden-Württemberg International Design Award

fo(u)r motion

Vierfach mobil
Fourfold mobility

Entwerfer / Developer
Ragnhild Albers
—
Hochschule / University
Fachhochschule Coburg
—
Betreuung / Support
Prof. Gerhard Kampe
Dipl.-Designerin Christiane Bausback
—

Zusammengeklappt zwei-, geöffnet vierrädrig und kombiniert mit geeigneten Komponenten, wird aus dem Fahrgestell von »fo(u)r motion« ein Wägelchen für kleine Lasten, eins für große, dann ein Fahrradanhänger und schließlich auch eine rollende Unterstützung für gangunsichere Personen. Das bietet nicht nur vielfältigeren Nutzen bei kleineren und größeren Unternehmungen, z.B. einkaufen, Veranstaltungen besuchen oder reisen; auch weitet sich der Nutzerkreis hin zu den Jüngeren.

—

Folded up it has two wheels, when opened out four, and when combined with appropriate components the chassis of »fo(u)r motion« is transformed into a small caddy for small loads, another for large loads, a bicycle trailer, and finally also a support on wheels for people who are unsteady on their feet. It not only offers more advantages for both modest and more ambitious projects, e.g. shopping, attending events and travel; it also expands the circle of users to include younger people.

—

Jury
Differenziertes Nutzungskonzept, modularer Aufbau und formale bzw. farbliche Ausgestaltung nehmen dem Gerät seinen ausgrenzenden Charakter. Die Feinarbeit an vielen Details, wie Klappmechanismus, Griffhöheneinstellung oder Gehstock, läßt breitere Akzeptanz und Tauglichkeit in Alter und Alltag erwarten. Statt einseitiger Gehhilfe ein variabler Gehilfe.

—

Judges panel
The varied usage concept, modular construction and the formal—or rather colour—design prevent »fo(u)r motion« from being restricted to a particular target group. Given the precision work on many details, such as the folding mechanism, the adjustment of the handlebar height or the walking stick, it is likely to be more widely accepted and generally more appropriate for elderly people. A variable, rather than a unilateral walking aid.

—

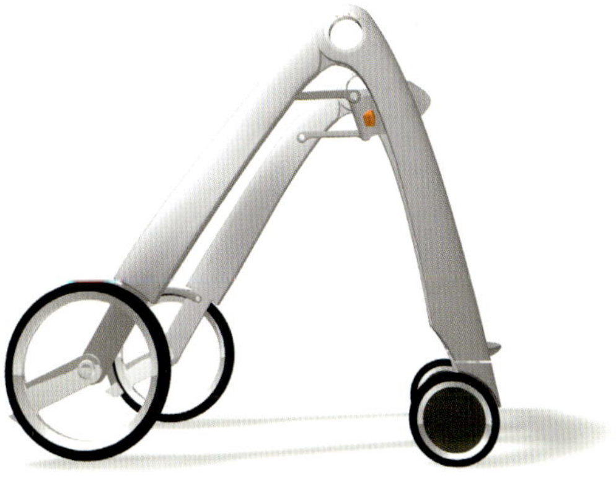

PUSH
PULL

Varus

Winkelgenau operieren
Precision-angle surgery

Entwerfer / Developer
Frank Fangmann
—
Hochschule / University
Fachhochschule Hannover
—
Betreuung / Support
Prof. Mike Nicklas
Dipl.-Designer Fred Held
—

Bestimmte Fehlstellungen des Hüftgelenks lassen sich durch Entnahme eines Keils aus dem Oberschenkel korrigieren. Bislang macht aber das genaue Sägen Probleme. Abhilfe schafft das System »Varus«: Am Knochen wird eine Winkellehre befestigt, die das oszillierende Blatt der Säge unter dem voreingestellten Winkel führt. Höhere Präzision, geringere Belastung für den Patienten, kürzere OP-Dauer, reduzierte Kosten sind die Folge.
—
Certain malpositions of the hip joint can be corrected by removing a wedge from the thigh. Until now, however, it has been difficult to saw with high precision. The »Varus« system remedies this situation: an angle gauge fastened to the bone guides the oscillating sawblade at the preset angle. The result is higher precision, less stress for the patient, shorter duration of surgery and lower costs.
—

Jury
Ein Problem wird einleuchtend analysiert und souverän gelöst. Die Gerätschaft ist deutlich und klar in ihre Funktionsbereiche Befestigung, Einstellung, Sägeführung und Bedienung gegliedert. Blaue Tefloneinsätze bzw. -ringe markieren die Essentials der Operationstechnik. Angemessene sachlich-technische Formensprache.
—
Judges panel
Logical analysis and solution of a problem—with consummate ease. The apparatus is clearly organized in its various functions: fastening, setting, saw guidance and control. Blue Teflon inserts or rings identify the essential features of the surgical technique. Appropriate formal expression, objectively and technically.
—

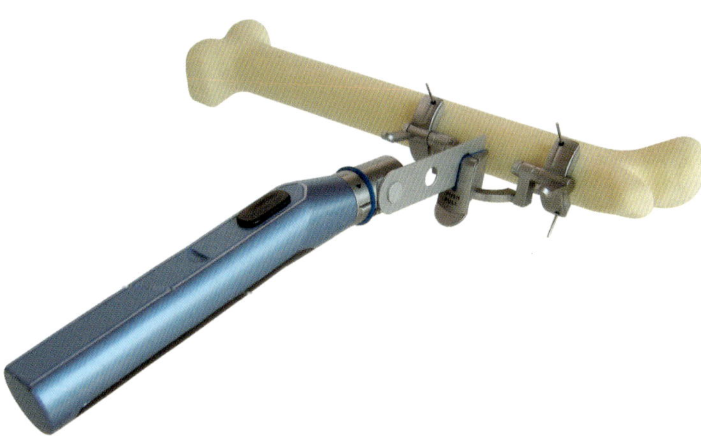

16:35
10-NOV-05
16:35
10-NOV-05

Senso

Erinnerungen wecken
Waking memories

Entwerfer / Developer
Pelin Celik
in Zusammenarbeit mit Pilotfish GmbH München
—
Hochschule / University
Fachhochschule Coburg
—
Betreuung / Support
Prof. Gerhard Kampe
Dipl.-Designer Harm Hogenbrink, Pilotfish GmbH
—

»Senso« ist eine Gruppe interaktiver Geräte, die dem Demenzkranken alle möglichen visuellen, akustischen, olfaktorischen und haptischen Informationen aus seiner Umwelt übermitteln, um an Vergessenes zu erinnern. Dazu trägt er um den Hals die Steuerungs- und Bedieneinheit »Personal Lifebook«, im Ohr einen »Souffleur« mit Kamera, Hörgerät und Sensoren für Vitalfunktionen und hat zu Hause ein »EMO Pillow«, das wohlig wärmt, pulsiert, leuchtet, Bilder und Filme zeigt, Düfte verströmt usw.
—
»Senso« is a group of interactive devices that convey every kind of visual, acoustic, olfactory and haptic information to dementia patients from their environment to remind them of things they have forgotten. To this end, patients wear a »Personal Lifebook« control and operating unit around their necks, have a »prompter« in their ear, with camera, hearing aid and sensors for vital functions, and at home an »EMO Pillow« that snugly warms them, pulsates, dispenses light, shows pictures and films, exudes fragrances etc.
—

Jury
Neueste Entwicklungen in Mikro-, Sensor-, Funk-, Material- und Energietechnologie sind aufgeboten, um dem Gedächtnisverlust auf- und dem Gefühl des Verlorenseins abzuhelfen. Szenarien von alltäglichen und bedrohlichen Situationen spielen das überzeugend durch. Die äußere Gestalt der Geräte ordnet sich dem Krankheitsgeschehen unter.
—
Judges panel
The latest developments in micro-, sensor, radio, materials and energy technology are called in to remedy loss of memory and the feeling of being lost. Scenarios of everyday and dangerous situations play it through convincingly. The outward form of the devices subordinates itself to the illness.
—

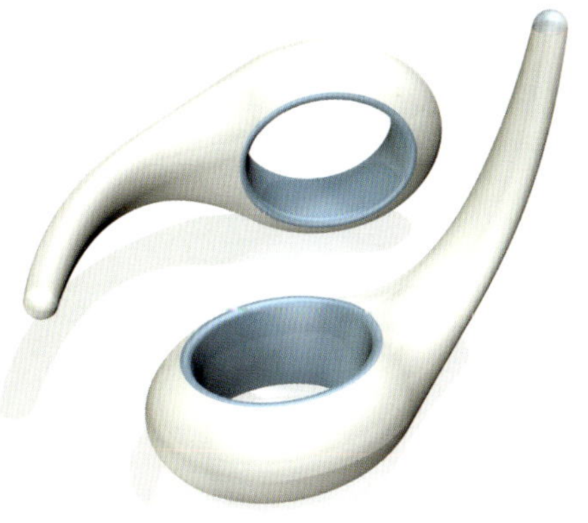

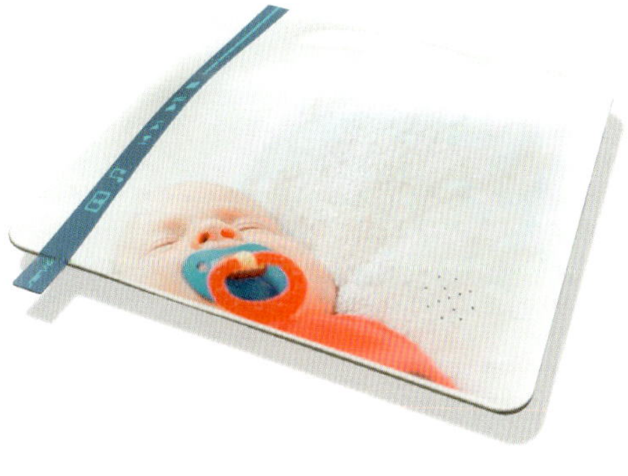

Lichthöhle

Schützende Raumblase
Protective space bubble

Entwerfer / Developer
Jens Krzywinski
–
Hochschule / University
Technische Universität Dresden
–
Betreuung / Support
Prof. Dr. Johannes Uhlmann, Dresden
Prof. Rolf Roeder, Halle
–

Die außenliegende pneumatische Struktur der »Lichthöhle« überspannt ein Quadrat von knapp 20 m^2, und zwar so, daß man überall stehen kann. Rund um das Oberlicht verläuft ein Infrastrukturring, der Kabel, Klimaleitungen, Leuchtmatten usw. aufnimmt und Staumöglichkeiten bietet; er wird zusammen mit dem Zelt zusammengefaltet. Ein-, Aus- bzw. Durchgänge an allen Seiten erlauben unterschiedliche Verbindungen der Zelte in Linie oder Karree.
–
The outer pneumatic structure of this »cave of light« covers a square area of almost 20 m^2 in such a way that it is possible to stand upright everywhere in it. An infrastructure ring surrounding the toplight carries cables, air-conditioning lines, lighting mats etc. and provides stowage space; it is folded up together with the tent. Entrances, exits or passages on all sides allow the tents to be joined together in rows or squares.
–

Jury
Die Konstruktion bedingt Features des Rettungszeltes, die dem Katastropheneinsatz besonders gut gerecht werden können. Mehr noch wiegt, dass der sich nach oben weitende Innenraum zusammen mit dem Verzicht auf Wandfenster eine Atmosphäre schafft, in der die hierher Verbrachten sich geborgen fühlen dürfen.
–
Judges panel
The design necessitates rescue tent features which could be particularly appropriate in emergency aid operations. What is more important is that the interior, which widens out towards the top, and the fact that there are no side windows create an atmosphere in which the people brought here can feel sheltered.
–

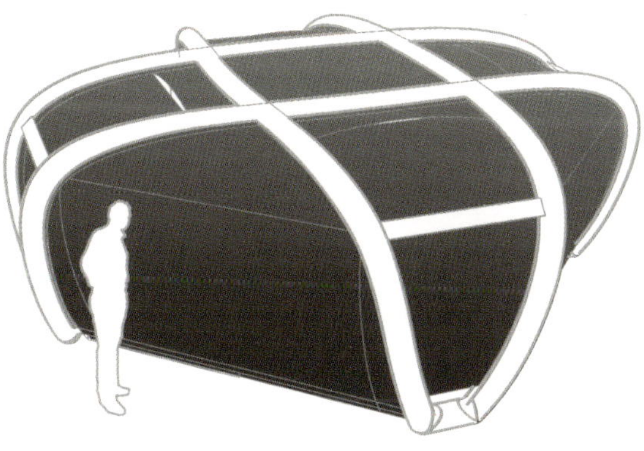

Tri-On

Kombinierter Schutz
Combined protection

Entwerfer / Developer
Julia Frank
–
Hochschule / University
Muthesius Kunsthochschule Kiel
–
Betreuung / Support
Prof. Ulrich Hirsch
–

Egal, in welcher Kombination Augen, Ohren und Atemwege bei der Arbeit zu schützen sind, immer wird ein universelles, einstellbares Kopfgestell getragen, daran sich nach Bedarf Brille, Ohrkapseln oder Maske anbringen lassen. Für jede Komponente ist auch eine Parkposition definiert. Dort, wo Dichtlippen für Formschluss zum Kopf sorgen sollen, ist individuelle Anfertigung nach 3D-Scannen des Gesichts vorgesehen.

–

In whatever combination eyes, ears and respiratory tracts have to be protected at work, a universal, adjustable head frame is always worn, to which goggles, ear plugs or a mask can be attached as needed. A docking position is also provided for each component. Sealing lips to ensure form-fit to the head are made to measure after performing a 3D scan of the face.

–

Jury

Die Arbeitsschutzausrüstung »Tri-On« bietet eine umfassende Lösung dafür, dass alle drei Schutzkomponenten auch wirklich gleichzeitig anwendbar sind, dabei optimal passen und außerdem einzeln und zusammen gut aussehen.

–

Judges panel

The »Tri-On« equipment for protection at work provides a complete solution: all three protective components really are usable simultaneously, they fit perfectly, and they look good both separately and together.

–

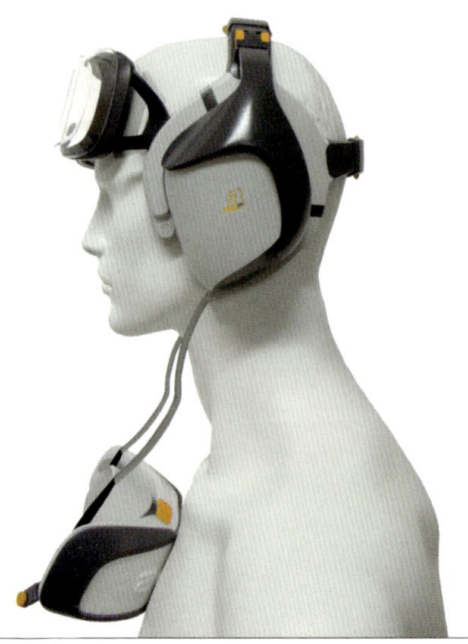

Bimo

Sozialer Arbeiter
Social worker

Entwerfer / Developer
Martin Postler
—
Hochschule / University
Muthesius Kunsthochschule Kiel
—
Betreuung / Support
Prof. Ulrich Hirsch
Prof. Dr. Norbert M. Schmitz
—

»Bimo« ist ein Roboter, der sich als Rad, das er in seiner Grundform ist, rollend oder auf sechs ausfahrbaren Beinen schrittweise bewegen kann. Die Energie für seine per Laserstrahl betätigten Polymer-Muskeln kommt aus einer Brennstoffzelle im Zentrum. Die Beine sind zugleich Arme, die sich aus einem Magazin mit Werkzeugen, Mess- oder medizinischen Instrumenten bestücken können. Die Idee ist, dass »Bimo« mit seinesgleichen kooperativ, arbeitsteilig und verständnisvoll zusammenwirkt und mit dem Menschen also ein »Metakollektiv« bildet.

—

»Bimo« is a robot that can move as a rolling wheel, which it is in its basic form, or step by step on six extendable legs. The energy for its laser beam-operated polymer muscles comes from a fuel cell at its centre. The legs also serve as arms, which can equip themselves with tools and measuring or medical instruments from a magazine. The idea is that »Bimo« should collaborate with other robots like itself, cooperating, sharing jobs and understanding, thus forming a »metacollective« with humans.

—

Jury
Gedacht ist hauptsächlich an den Einsatz bei Naturkatastrophen, Epidemien oder verheerenden Unfällen, aufgrund der Multifunktionalität aber auch an die Assistenz im sozialen und privaten Leben. Umfangreiche Überlegungen aus Robotik, Kybernetik und KI untermauern den zukunftsweisenden Entwurf.

—

Judges panel
The main area of use envisaged is in natural disasters, epidemics or devastating accidents, due to its multifunctionality, though, also as an assistant in social work or private life. Numerous considerations derived from robotics, cybernetics and AI support the forward-looking design.

—

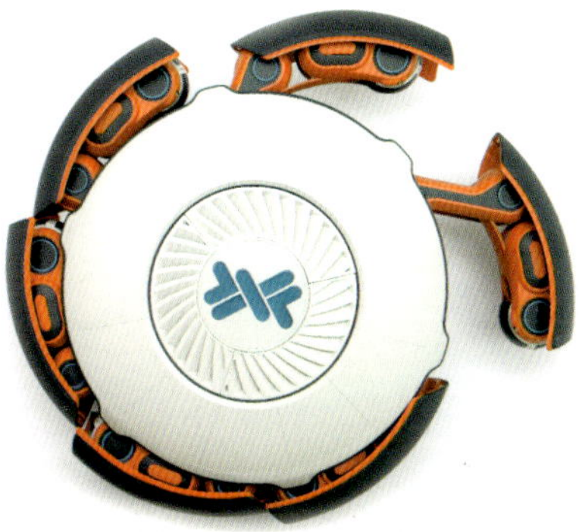

Coaster

Geländegängiger Rollstuhl
Off-road wheelchair

Entwerfer / Developer
Peer Stock
—
Hochschule / University
Kunsthochschule Berlin-Weißensee
—
Betreuung / Support
Prof. Dietmar Palloks
—

»Coaster« soll dem sportlichen Gehbehinderten etwas bieten, das dem Mountainbike nahe kommt: ein geländefähiges, vierrädriges Fahrzeug mit Hebelantrieb und Nabenschaltung. Gelenkt und gebremst wird mit Hebeldrehung. Ein schwenkbarer Schalensitz ermöglicht einen einfachen Einstieg. Carbonfaserverbundwerkstoffe und Leichtmetalle halten das Gewicht gering.

—

»Coaster« is intended to offer something similar to a mountain-bike for sports enthusiasts with a mobility handicap: a four-wheeled all-terrain vehicle with lever drive and hub gears. It is steered and braked by rotating levers. A swivelling bucket seat makes it easy to get into the wheelchair. Carbon fibre composites and light alloys keep the weight down.

—

Jury
Das Konzept eröffnet Gehbehinderten ein neues Feld sportlicher Aktivitäten. Sinnvollerweise wird bei Funktionen wie Bremsen, Federn oder Schalten auf vorhandene Komponenten der Fahrradtechnik zurückgegriffen. Für Probleme wie Kraftübertragung und Lenkung werden eigene, neue Lösungen vorgeschlagen.

—

Judges panel
The concept opens up a new area of sporting activities for people with a mobility handicap. Sensibly, existing cycle components are used for functions such as braking, suspension and changing gears. New, original solutions are proposed for problems such as power transmission and steering.

—

Aerius

Mobile Beatmung
Mobile respiration

Entwerfer / Developer
Grit Andrea Wolf
—
Hochschule / University
Hochschule Anhalt (FH)
—
Betreuung / Support
Prof. Uwe Gellert
Dipl.-Formgestalter Klaus Heller
—

Eine minimalinvasive Beatmungsmethode macht den Patienten mit Atmungsinsuffizienz wenigstens in Wohnung und Wohnumfeld mobil. »Aerius« enthält in seinem bereiften Zylinder die erforderlichen Aggregate: Kompressor, Akkumulator und die herausnehmbare Luftbefeuchtereinheit. Der höhenverstellbare Griffstab führt auch den Luftschlauch zum Patienten. Die Ladestation ist zugleich Parkplatz. Anderswo steht das Gerät wie ein Stehaufmännchen.

—

A minimally invasive respiration method enables patients with respiratory insufficiency to move around, at least at or near home. In its wheeled cylinder, »Aerius« contains the necessary units: compressor, battery and the removable air humidifier unit. The height-adjustable handlebar also carries the air line to the patient. The recharging point doubles as a parking space. Elsewhere, the apparatus is stood upright.

—

Jury
Das Mobilitätskonzept erspart dem Patienten den stationären Klinikaufenthalt und gibt ihm so mehr Selbständigkeit. Unter medizinischer Überwachung und mit vernünftigem Aufwand kann er das Gerät in seinen Alltag einbeziehen – und gewinnt erheblich an Lebensqualität.
—
Judges panel
The mobility concept makes hospitalization unnecessary, thus giving patients more independence. Under medical supervision they can integrate the device in their everyday lives at reasonable cost—with a substantial improvement in their quality of life.
—

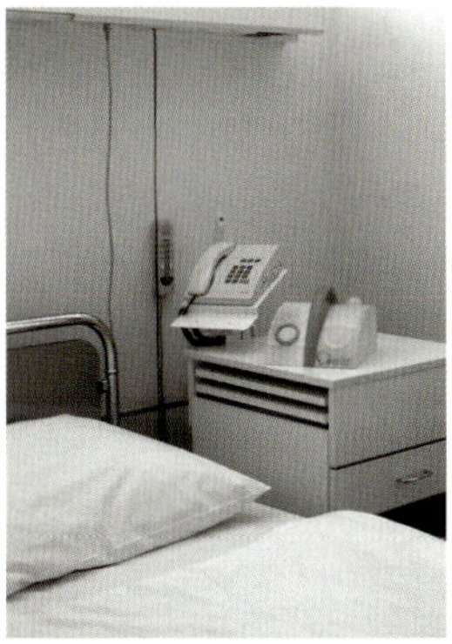

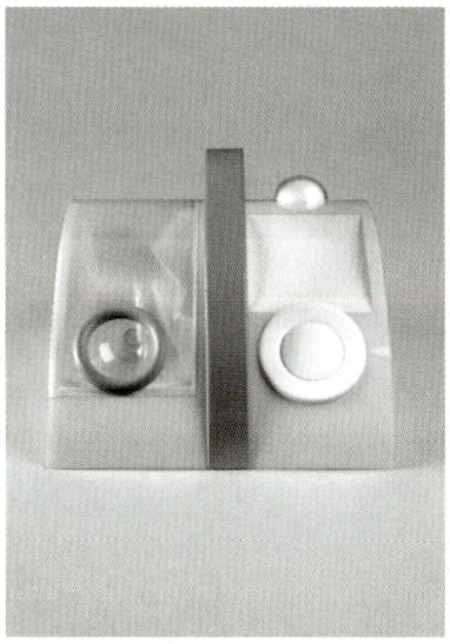

Siento

Musik für Gehörlose
Music for the deaf

Entwerfer / Developer
Ufuk Yüksel
—
Hochschule / University
Fachhochschule Coburg
—
Betreuung / Support
Prof. Gerhard Kampe
—

»Siento« (span.: ich fühle) macht Musik über Vibrationen am Körper wahrnehmbar. Das um den Hals getragene Hauptgerät rechnet die Musikinformation um und gibt sie über die beiden sog. Exciter in den Ausläufern an die Haut weiter. Ergänzend erhalten die beiden Satellitengeräte, ein Armband und ein Handschmeichler, dieselben Informationen per Funk.

—

»Siento« (Spanish for »I feel«) makes music perceptible via vibrations on the body. The main unit, worn around the neck, converts the music data and transmits it to the skin via the two exciters in the extensions. Complementing this, the two satellite units, a bracelet and a hand stroker, receive the same data wirelessly.

—

Jury
Über Intensität und Dauer der Impulse und abhängig davon, wo die Reizung stattfindet, werden charakteristische Eigenschaften des gespielten Musikstücks erlebbar, auch von Gehörlosen und Hörenden gemeinsam. In diesem ästhetischen Zusammenhang ist es nur konsequent, die Schmuckfunktion der Gehäuseformen zu betonen.
—
Judges panel
Through the intensity and duration of the stimuli, and depending on where the stimulation takes place, characteristic qualities of the piece of music being played can be experienced—also by deaf and hearing people together. In this aesthetic context it is only logical to emphasize the decorative function of the housing designs.
—

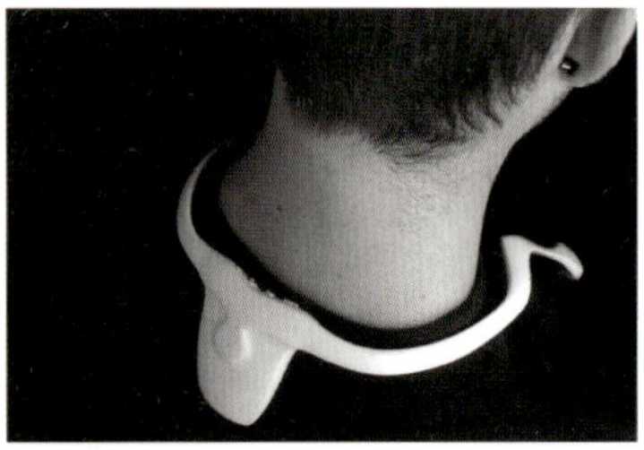

Mia Seeger Stiftung
Mia Seeger Foundation

Mia Seeger (1903 – 1991)

Mia Seeger war die »Grande Dame« des Design. Vor dem Zweiten Weltkrieg wirkte sie an großen Ausstellungsprojekten des Deutschen Werkbundes mit, 1927 z.B. an der Weißenhofsiedlung in Stuttgart. Nach dem Krieg führte sie als Präsidialmitglied zwölf Jahre lang den Rat für Formgebung in Darmstadt. 1986 rief sie die nach ihr benannte Stiftung ins Leben, deren Zweck die umfassende Förderung des Design als Teil von Wirtschaft und Kultur ist. Namhafte Sponsoren aus der Wirtschaft haben sich ihren Zielen angeschlossen.

Mit der Absicht, besonders den Nachwuchs im Design zu fördern und ihn dabei zur Auseinandersetzung mit sozialen Fragen herauszufordern, schreibt die Stiftung jährlich den Mia Seeger Preis bundesweit aus, zusätzlich den Mia Seeger Förderpreis zur Unterstützung von Projekten.

Ausführliche Informationen zur Stiftung über Internet www.mia-seeger.de oder bei der Geschäftsstelle.

Mia Seeger Stiftung
c/o Design Center Stuttgart
im Haus der Wirtschaft
Willi-Bleicher-Straße 19
70174 Stuttgart
T +49 7 11 1 23 26 84 oder 26 38
F +49 7 11 1 23 25 77
miaseegerstiftung@web.de

Impressum
Herausgeber: Mia Seeger Stiftung
Redaktion: Wolfgang Berger
Grafikdesign: Stapelberg & Fritz
Ausstellungsgestaltung: design hoch drei
Plakat: Uwe Lohrer
Fotos: Preisträger und Ausgezeichnete

Stuttgart 2006

Mia Seeger was the »Grande Dame« of design. Before the Second World War she was involved in major exhibition projects of the Deutscher Werkbund, such as the Weissenhof complex in Stuttgart in 1927. After the war she sat for twelve years on the managing committee of the Rat für Formgebung (Design Council) in Darmstadt. In 1986, she established the foundation that bears her name, whose purpose is the comprehensive promotion of design as a part of economic and cultural life. Prominent sponsors from the business world have lent their support to her aims.

With the aim of promoting young designers in particular, and providing them with the additional challenge of coming to terms with social problems, the Foundation invites entrants from all over Germany to compete for the annual Mia Seeger Prize. There is also a Mia Seeger Scholarship to support projects.

For more details about the Foundation, visit www.mia-seeger.de or contact the Foundation office.

Mia Seeger Stiftung
c/o Design Center Stuttgart
im Haus der Wirtschaft
Willi-Bleicher-Strasse 19
70174 Stuttgart
T +49 7 11 1 23 26 84 or 26 38
F +49 7 11 1 23 25 77
miaseegerstiftung@web.de

Publishing details
Published by: Mia Seeger Stiftung
Editor: Wolfgang Berger
Graphic design: Stapelberg & Fritz
Exhibition design: design hoch drei
Poster: Uwe Lohrer
Photographs: prize-winners and commended entrants

Stuttgart 2006

Adressen
Addresses

Adam + Partner
Chris Adam
Bahnhofstraße 13
D-82467 Garmisch
T +49 1 75 5 28 96 24
F +49 1 80 50 60 34 88 79 69
scda1@t-online.de
Seite / Page 56
—

AEG Elektrowerkzeuge
Max-Eyth-Straße 10
D-71364 Winnenden
T +49 71 95 1 20
F +49 71 95 1 26 66
Volker.Siegle@AEG-Milwaukee.com
www.AEG-PT.com
Seite / Page 44
—

Ahape – more than design
Ahape GmbH
Eckhard Ruoff und Herbert Pauser
Elisabethenstraße 34
D-64283 Darmstadt
T +49 61 51 5 01 23 50
F +49 61 51 5 01 23 51
info@ahape.de
www.ahape.de
Seite / Page 46
—

Ragnhild Albers
Deisenhofenerstraße 59
D-81539 München
Mobil +49 1 72 5 34 56 20
ragnhildalbers@gmx.de
Seite / Page 216
—

Atelier für Gestaltung Papenfuss
Marktstraße 12
D-99423 Weimar
T +49 36 43 7 76 00
F +49 36 43 77 60 20
mail@atelierpapenfuss.de
www.atelierpapenfuss.de
Seite / Page 152
—

Auerhahn Bestecke GmbH
Im oberen Tal 9
D-72213 Altensteig
T +49 74 53 94 68 34
F +49 74 53 94 68 93
info@auerhahn-bestecke.de
www.auerhahn-bestecke.de
Seite / Page 106
—

Jacob Bachmayer
Nordbahnhofstraße 191
D-70191 Stuttgart
T +49 7 11 8 82 04 98
F +49 7 11 81 04 08 31
jacob@renschler.de
Seite / Page 198
—

Mario Bellini mit Claudio Bellini
c/o Vitra GmbH
Charles-Eames-Straße 2
D-79576 Weil am Rhein
T +49 76 21 70 20
F +49 76 21 7 02 32 42
Seite / Page 130
—

Bessey Tool GmbH & Co. KG
Postfach 1154
D-74301 Bietigheim-Bissingen
T +49 71 42 40 10
F +49 71 42 40 14 51
info@bessey.de
www.bessey.de
Seite / Page 48
—

bgmbh Medien + Datenbanken
Kurfürstenstraße 22
D-71636 Ludwigsburg
T +49 71 41 6 88 92 00
F +49 71 41 6 88 92 17
info@bgmbh.de
www.bgmbh.de
Seite / Page 64
—

BGP Design
Lindenstraße 10
D-70563 Stuttgart
T +49 7 11 22 06 28 90
F +49 7 11 2 20 62 89 99
braake@bgp-design.com
www.bgp-design.com
Seite / Page 50
—

Jörg Boner
Edenstraße 16
CH-8045 Zürich
T +41 44 2 01 79 34
F +41 44 2 01 79 35
mail@joergboner.ch
www.joergboner.ch
Seite / Page 138
—

Robert Bosch Hausgeräte GmbH
Carl-Wery-Straße 34
D-81739 München
T +49 89 45 90 02
F +49 89 45 90 23 47
martina.schellong@bshg.com
www.bosch-hausgeraete.de
Seite / Page 80, 86, 88, 114

Werksdesign / In-house design
Alexander Marsch
Seite / Page 88
und
Roland Vetter
Seite / Page 80, 86
—

BREE Collection GmbH & Co. KG
Gerberstraße 3
D-30916 Isernhagen
T +49 51 36 8 97 67 03
F +49 51 36 8 97 67 05
bree.collection@bree.de
www.bree.de

BREE Design Team
Tim Knossalla
Seite / Page 202, 204
—

Jürgen Breitkopf Products
Nottulner Landweg 104
D-48161 Münster
T +49 25 34 64 58 66
info@breitkopf-products.de
www.breitkopf-products.de
Seite / Page 60
—

Calimax GmbH
Bundesstraße 102
A-6830 Rankweil
T +43 55 22 8 36 77
F +43 55 22 83 67 76
office@calimax.com
www.calimax.com
Seite / Page 34
—

Pelin Celik
Konrad-Adenauer-Straße 14
D-78549 Spaichingen
Mobil +49 1 60 91 13 22 85
celikp@web.de
Seite / Page 220
—

Clage GmbH
Pirolweg 1–5
D-21337 Lüneburg
T +49 41 31 8 90 10
F +49 41 31 8 32 00
info@clage.de
www.clage.de
Seite / Page 78
—

Conergy AG
Anckelmannsplatz 1
D-20537 Hamburg
T +49 40 2 37 10 20
F +49 40 2 37 10 21 48
info@conergy.de
www.conergy.de

Werksdesign / In-house design
Sven Dähne, Peer Stock, Martin Dorsch
Seite / Page 24
—

DaimlerChrysler AG
Epplestraße 225
D-70546 Stuttgart
www. daimlerchrysler.com

T +49 7 11 1 70
F +49 7 11 1 72 22 44
www.mercedes-benz.de
Seite / Page 184

T +49 7 11 1 79 52 51
matthias.steybe@daimlerchrysler.com
Seite / Page 66

T +49 7 11 1 79 72 60
benjamin.syring@daimlerchrysler.com
Seite / Page 68
—

DaimlerChrysler
Design Team
Benzstraße X 800
D-71063 Sindelfingen
T +49 70 31 9 07 96 25
F +49 70 31 9 07 98 90
Seite / Page 184
—

DaimlerChrysler
Vertriebsorganisation Berlin
D-10878 Berlin
T +49 30 2 69 40
F +49 30 26 94 39 99
Seite / Page 184
—

Daniels & Koitzsch
Bessungerstraße 12–14
D-64285 Darmstadt
T +49 61 51 9 51 66 80
F +49 61 51 9 51 69 00
design@daniels-koitzsch.de
www.daniels-koitzsch.de
Seite / Page 114
—

Design Atelier Maurer GmbH
Fabrik am Rotbach / Eisenhalle
CH-9055 Bühler
T +41 71 7 90 08 60
F +41 71 7 90 08 60
info@waltermaurer.ch
www.waltermaurer.ch

Werksdesign / In-house design
Walter Maurer
Seite / Page 206
—

Design hoch drei GmbH & Co. KG
Hallstraße 25a
D-70376 Stuttgart
T +49 7 11 55 03 77 30
F +49 7 11 55 03 77 55
info@design-hoch-drei.de
www.design-hoch-drei.de
Seite / Page 66, 68
—

Adressen
Addresses

Design Planet
José Delhaes
Hausener Straße 2
D-63165 Mühlheim am Main
T +49 61 08 7 96 70
F +49 61 08 79 67 20
info@designplanet.de
www.designplanet.de
Seite / Page 78
—

Designpraxis Diener
Irisweg 3
D-89079 Ulm
T +49 73 05 9 64 80
F +49 73 05 96 48 30
info@designpraxis-diener.de
www.designpraxis-diener.de
Seite / Page 54
—

Design Tech
Jürgen R. Schmid
Zeppelinstraße 53
D-72119 Ammerbuch
T +49 70 73 9 18 90
F +49 70 73 91 89 17
info@designtechschmid.de
www.designtechschmid.de
Seite / Page 34, 74, 168
—

Design-Werkstatt Weizenkorn
Binningerstraße 97a
CH-4123 Allschwil
T +41 61 4 81 40 00
F +41 61 4 81 60 16
design-werkstatt@weizenkorn.ch
www.weizenkorn.ch
Seite / Page 62
—

Stefan Diez
Geyerstraße 20
D-80469 München
T +49 89 20 24 53 92
F +49 89 20 24 53 95
mail@stefan-diez.com
www.stefan-diez.com
Seite / Page 142
—

DNS Designteam
Tübinger Straße 92
D-70178 Stuttgart
T +49 7 11 6 40 71 86
F +49 7 11 6 40 71 88
stuttgart@dns-design.de
www.waterbelt.com
www.dns-design.de

Werksdesign / In-house design
Ralph Schäflein + Kai-Uwe Neth
Seite / Page 192
—

Duravit AG
Werderstraße 36
D-78132 Hornberg
T +49 78 33 7 00
F +49 78 33 7 02 89
duravit@duravit.de
www.duravit.de
Seite / Page 76
—

Albert Ebenbichler
Innere Brücke 4
D-73728 Esslingen
T +49 7 11 3 51 12 80
F +49 8 91 4 88 17 07 11
info@albert-ebenbichler.com
www.albert-ebenbichler.com
Seite / Page 148
—

Eisele Kuberg Design
Frank Eisele
Oderstraße 1
D-89231 Neu-Ulm
T +49 7 31 9 80 75 55
F +49 7 31 9 80 75 56
info@eiselekubergdesign.de
www.eiselekubergdesign.de
Seite / Page 86, 194
—

Eisfink Max Maier GmbH & Co. KG
Rheinlandstraße 11
D-71636 Ludwigsburg
T +49 71 41 47 91 45
F +49 71 41 47 92 00
info@eisfink.de
www.eisfink.de
Seite / Page 116, 118
—

Hartmut S. Engel
Design Studio
Monrepos-Straße 7
D-71634 Ludwigsburg
T +49 71 41 2 34 50
F +49 71 41 23 45 20
h.s.engel@designstudioengel.de
Seite / Page 170, 172, 174
—

EOOS Design GmbH
Zelinkagasse 2/6
A-1010 Wien
T +43 1 4 05 39 87
F +43 1 4 05 39 87 80
design@eoos.com
www.eoos.at
Seite / Page 132
—

Erco Leuchten GmbH
Brockhauser Weg 80–82
D-58507 Lüdenscheid
T +49 23 51 55 14 55
F +49 23 51 55 16 75
h.kosche@erco.com
www.erco.com

Werksdesign / In-house design
Henk Kosche, Alois Dworschak
Seite / Page 166
—

Eternit AG
CH-8867 Niederurnen
T +41 55 6 17 13 68
F +41 55 6 17 14 73
daniel.hauri@eternit.ch
www.eternit.ch
Seite / Page 200
—

Eva Denmark A/S
Måløv Teknikerby 18–20
DK-2760 Måløv
T +45 36 73 20 60
F +45 36 70 74 11
mail@evadenmark.com
www.evadenmark.com
www.evasolo.com
Seite / Page 96, 110, 112
—

Eva Denmark
Vertriebsleitung Deutschland
Thomas Schmidt
Nordallee 17
D-56112 Lahnstein
T +49 26 21 62 96 03
F +49 26 21 18 76 10
tsc@evadenmark.com
Seite / Page 96, 110, 112
—

Frank Fangmann
Große Barlinge 8
D-30171 Hannover
Mobil +49 1 77 7 37 35 56
frank.fangmann@gmx.de
Seite / Page 218
—

Fissler GmbH
Harald-Fissler-Straße 1
D-55743 Idar-Oberstein
T +49 67 81 40 30
F +49 67 81 40 30
produktmarketing@fissler.de
www.fissler.de
Seite / Page 104
—

Julia Frank
Am Redder 34
D-25336 Klein Nordende
Mobil +49 1 79 7 47 08 62
juliafrank.kiel@gmx.de
Seite / Page 224
—

Patrick Frey + Markus Boge
Leinaustraße 6
D-30451 Hannover
T +49 5 11 3 37 87 91
F +49 5 11 3 37 87 92
mail@frey-boge.de
www.frey-boge.de
Seite / Page 140
—

Jan-Hendrik de Groote und Vincent Jalet
Pierre Corneliskaai 35
B-9300 Aalst
T +32 53 72 75 46
F +32 53 72 75 40
Seite / Page 100, 102
—

Hadi Teherani AG
Hadi Teherani, Hans Ulrich Bitsch, Ulrich Nether
Oberbaumbrücke 1
D-20427 Hamburg
T +49 40 4 44 05 43
F +49 40 44 40 54 43
info@haditeherani.de
www.haditeherani.de
Seite / Page 128
—

Handelsagentur CDH
Dorfackerstraße 6
D-72074 Tübingen
T +49 70 71 25 21 88
F +49 70 71 25 21 99
Seite / Page 206
—

Windmühlenmesser
Robert Herder GmbH & Co. KG
Ellerstraße 16
D-42697 Solingen
T +49 2 12 26 70 50
F +49 2 12 7 53 27
info@windmuehlenmesser.de
www.windmuehlenmesser.de

Werksdesign / In-house design
Giselheid Herder-Scholz
Seite / Page 108
—

Hess Form + Licht GmbH
Schlachthausstraße 19–19/3
D-78050 Villingen-Schwenningen
T +49 77 21 92 00
F +49 77 21 92 02 50
hess@hess-form-licht.de
www.hess-form-licht.de
Seite / Page 162, 164
—

Industrial PDD GmbH & Co. KG
Calwer Straße 11
D-70173 Stuttgart
T +49 7 11 6 15 86 20
F +49 7 11 6 15 86 21
mail@industrial.li
www.industrial.li
Seite / Page 72, 176
—

Interstuhl Büromöbel GmbH & Co. KG
Brühlstraße 21
D-72469 Meßstetten-Tieringen
T +49 74 36 87 10
F +49 74 36 87 13 13
info@interstuhl.de
www.interstuhl.de
www.silverchair.net
Seite / Page 128
—

Adressen
Addresses

Interstuhl Büromöbel GmbH & Co. KG
Vertrieb
Brühlstraße 21
D-72469 Meßstetten-Tieringen
T +49 74 36 87 10
F +49 74 36 87 11 10
Seite / Page 128
—

James Irvine
Via Sirtori 4
I-20129 Mailand
T +39 2 29 53 45 32
F +39 2 29 53 45 31
james@james-irvine.com
www.james-irvine.com
Seite / Page 146
—

Claus Jensen & Henrik Holbæk
Tools®
Rentemestervej 23 A
DK- 2400 Kopenhagen NV
T +45 38 19 41 14
F +45 38 19 41 13
email@toolsdesign.dk
www.toolsdesign.dk
Seite / Page 96, 110, 112
—

Job Factory AG
Bordeaux-Straße 5
CH-4057 Basel
T +41 61 5 60 01 01
F +41 61 5 60 01 11
ic@jobfactory.ch
www.jobfactory.ch
Seite / Page 62
—

Alfred Kärcher GmbH & Co. KG
Alfred-Kärcher-Straße 28–40
D-71364 Winnenden
T +49 71 95 1 40
F +49 71 95 14 22 12
info@kaercher.de
www.kaercher.de

Werksdesign / In-house design
Denis Dammköhler
Seite / Page 52
und
Michael Meyer
Seite / Page 42
—

Kesseböhmer GmbH
Mindener Straße 208
D-49152 Bad Essen
T +49 57 42 46 13 04
F +49 57 42 4 66 13 04
h.isfort@kesseboehmer.de
www.kesseboehmer.de
Seite / Page 122
—

Kienledesign
Staufenstraße 9
D-73575 Leinzell
T +49 71 75 91 91 15
F +49 71 75 91 91 14
info@kienledesign.de
www.kienledesign.de
Seite / Page 44
—

KPG Design Group GmbH
Leinzellerstraße 14
D-73527 Täferrot
T +49 71 75 9 99 12 30
F +49 71 75 9 99 12 20
georg.kunz@kpg.de
www.kpg.de
Seite / Page 26, 190
—

Jens Krzywinski
Gostritzer Straße 12
D-01217 Dresden
Mobil +49 1 79 9 28 09 57
jens.krzy@web.de
Seite / Page 222
—

C. Josef Lamy GmbH
Grenzhöfer Weg 32
D-69123 Heidelberg
T +49 62 21 84 31 04
F +49 62 21 84 31 21
wolfgang.weber@lamy.de
www.lamy.de
Seite / Page 150
—

Leise®
Designentwicklung und Vertrieb
Töngesgasse 40
D-60311 Frankfurt am Main
T +49 69 21 99 98 01
F +49 69 21 99 98 02
contact@leise-leise.com
www.leise-leise.com
Seite / Page 136
—

Leise Studio
Knut Völzke
Töngesgasse 40
D-60311 Frankfurt am Main
T +49 69 21 99 98 01
F +49 69 21 99 98 02
k.voelze@leise-leise.com
www.leise-leise.com
Seite / Page 136
—

Maquet GmbH
Kehler Straße 31
D-76437 Rastatt
T +49 72 22 93 23 28
F +49 72 22 93 28 28
info.magnus@maquet.com
www.maquet.com
www.maquet-magnus.de
Seite / Page 74
—

Maquet Vertrieb & Service Deutschland GmbH
Kehler Straße 31
D-76437 Rastatt
T +49 72 22 93 25 90
F +49 72 22 93 29 81
info.vertrieb@maquet.de
www.maquet.de
Seite / Page 74
—

Mayer und Thiele GbR
Treppenstraße 17
D-42115 Wuppertal
T +49 2 02 4 46 98 02
F +49 2 02 4 46 98 01
office@mayerundthiele.de
www.mayerundthiele.de
Seite / Page 98
—

Mehnert Corporate Design GmbH
Köpenicker Straße 48–49
D-10179 Berlin
T +49 30 6 11 11 99
F +49 30 6 11 11 98
je@mehnertdesign.de
www.mehnertdesign.de
Seite / Page 38
—

Metz & Kindler Produktdesign
Frankfurter Straße 44
Im Kontorhaus
D-64293 Darmstadt
T +49 61 51 29 36 41
F +49 61 51 29 52 52
public@metz-kindler.de
www.metz-kindler.de
Seite / Page 84
—

Miele & Cie. KG
Carl-Miele-Straße 29
D-33332 Gütersloh
T +49 52 41 8 90
F +49 52 41 89 20 90
info@miele.de
www.miele.de
Seite / Page 92
—

Milani design + consulting
Seestraße 78
CH-8703 Erlenbach-Zürich
T +41 44 9 14 74 74
F +41 44 9 14 74 75
info@milani.ch
www.milani.ch
Seite / Page 104
—

Modus Product Design GmbH
Markgröninger Straße 61–1
D-70435 Stuttgart
T +49 7 11 8 70 21 05
F +49 7 11 8 70 21 06
office@modusdesign.de
www.modusdesign.de
Seite / Page 30
—

Nils Holger Moormann GmbH
An der Festhalle 2
D-83229 Aschau im Chiemgau
T +49 80 52 9 04 50
F +49 80 52 90 45 45
info@moormann.de
www.moormann.de
Seite / Page 138, 140
—

Martin Mostböck
Schönlaterngasse 5/4/3
A-1010 Wien
T +43 1 5 13 65 90
F +43 1 5 13 65 90
mail@martin-mostboeck.com
www.martin-mostboeck.com
Seite / Page 200
—

Pauli's Tauchshop
Hauptstraße 119
D-89250 Senden
T +49 73 07 80 09 00
F +49 73 07 80 09 50
Seite / Page 194
—

Paulmann Licht GmbH
Quezinger 2
D-31832 Springe
T +49 50 41 99 82 36
F +49 50 41 9 98 42 36
mvogt@paulmann.de
www.paulmann.de

Werksdesign / In-house design
Matthias Vogt
Seite / Page 178
—

Adressen
Addresses

Pearl Creative
Tim Storti, Christian Rummel
und Max Maier
Rheinlandstraße 10
D-71636 Ludwigsburg
T +49 71 41 4 88 74 90
F +49 71 41 4 88 74 99
info@pearlcreative.com
www.pearlcreative.com
Seite / Page 116, 118, 120
—

Joachim Pflug
Robert-Bosch-Straße 42
D-72810 Gomaringen
T +49 70 72 21 44
F +49 70 72 8 06 57
info@pflug-innenausbau.de
www.pflug-innenausbau.de
Seite / Page 32
—

Phoenix Design
Kölner Straße 16
D-70376 Stuttgart
T +49 7 11 9 55 97 60
F +49 7 11 55 93 92
Andreas.Haug@phoenixdesign.de
www.phoenixdesign.de
Seite / Page 28, 36
—

Martin Postler
Longstreet Workshops
Building 7, Unit 8
GB-London E2 8HN
T +44 79 46 94 64 51
martinpostler@gmx.de
martin.postler@rca.ac.uk
Seite / Page 225
—

Renschler
Kirchtalstraße 30
D-70435 Stuttgart
T +49 7 11 87 94 62
F +49 7 11 8 70 11 80
info@renschler.de
www.renschler.de
Seite / Page 198
—

Rieber GmbH & Co. KG
Hoffmannstraße 44
D-72770 Reutlingen
T +49 71 21 51 80
F +49 71 21 51 83 02
info@rieber.de
www.rieber.de
Seite / Page 120
—

Rosenbauer Deutschland
Carl-Metz-Straße 3
D-76185 Karlsruhe
T +49 7 21 5 96 50
F +49 7 21 5 96 51 72
Seite / Page 182
—

Rosenbauer International AG
Paschinger Straße 90
A-4060 Leonding
T +43 7 32 6 79 40
F +43 7 32 67 94 89
gerda.koenigstorfer@rosenbauer.com
www.rosenbauer.com
Seite / Page 182
—

Rosenthal Studio-Line
Rosenthal AG
Philip-Rosenthal-Platz 1
D-95100 Selb
T +49 92 87 7 25 66
F +49 92 87 7 22 71
beate.reichel@rosenthal.de
www.rosenthal.de

Design / Designer
Little Wonder
Gyungju Chyon, John Stanislav Sadar
Seite / Page 144
—

Rotinor GmbH
Flachter Straße 32
D-70499 Stuttgart
T +49 7 11 9 93 39 70
F +49 7 11 99 33 97 33
grimm@rotinor.com
www.seabob.com
Seite / Page 196
—

Rotinor
Marketing
Eckendorfer Straße 32
D-33609 Bielefeld
T +49 5 21 9 31 12 82
F +49 5 21 9 31 12 88
info@seabob.com
Seite / Page 196
—

Schefenacker AG
Alfred-Schefenacker-Straße 1
D-71409 Schwaikheim
T +49 71 95 5 81 11 14
F +49 71 95 5 81 13 76
alf.liesener@schefenacker.com

Werksdesign / In-house design
Alf Liesener
Seite / Page 186
—

Schmiddem
Ansbacher Straße 71
D-10777 Berlin
T +49 30 61 40 16 24
F +49 30 61 40 16 26
team@schmiddem-design.de
www.schmiddem-design.de
Seite / Page 76
—

Schnabel, Schneider Industrial Design
Weilerstraße 14
D-73614 Schorndorf
T +49 71 81 9 80 28 88
F +49 71 81 9 80 28 82
info@schnabel-schneider.de
www.schnabel-schneider.de
Seite / Page 64
—

Schneider & Held GmbH
Breite Straße 3–1
D-88481 Balzheim
T +49 73 47 92 11 01
F +49 73 47 91 96 13
Seite / Page 200
—

Schneider Novus Vertriebs GmbH
Holzhofring 20
D-82362 Weilheim
T +49 8 81 62 90
F +49 8 81 6 29 33
Seite / Page 152
—

Schneider Schreibgeräte GmbH
Am Schwarzenbach
D-78144 Tennenbronn
T +49 77 29 88 80
F +49 77 29 8 88 88
info@schneidermail.de
www.schneiderpen.de
Seite / Page 152
—

Schweizer Design Consulting
Heusteigstraße 19
D-70182 Stuttgart
T +49 7 11 8 06 97 70
F +49 7 11 80 60 72 74
info@schweizerdesign.com
www.schweizerdesign.com
Seite / Page 196
—

Siemens Electrogeräte GmbH
Carl-Wery-Straße 34
D-81739 München
T +49 89 45 90 09
F +49 89 45 90 23 47
elfriede.engelmann@bshg.com
www.siemens-hausgeraete.de

Werksdesign / In-house design
Gerd G. Wilsdorf,
Frank Rieser, Heiko Thielen
Seite / Page 82, 90
—

Sigma Elektro GmbH
Dr. Julius-Leber-Straße 15
D-67433 Neustadt
T +49 63 21 9 12 00
F +49 63 21 9 12 34
info@sigmasport.com
www.sigmasport.de
Seite / Page 176
—

SMA Technologie AG
Hannoversche Straße 1–5
D-34266 Niestetal
T +49 5 61 9 52 20
F +49 5 61 9 52 21 00
info@SMA.de
www.SMA.de
Seite / Page 30
—

Spirit Design
Silbergasse 8
A-1190 Wien
T +43 1 3 67 79 79 0
F +43 1 3 67 79 79 70
spirit@spiritdesign.at
www.spiritdesign.at
Seite / Page 182
—

Peer Stock
Eggerstedtstraße 80
D-22765 Hamburg
Mobil +49 1 63 7 35 50 21
peer.stock@gmx.de
Seite / Page 24, 226
—

Andreas Störiko
Corso Magenta 32
I-20123 Mailand
T +39 02 89 01 29 71
storiko@compuserve.com
Seite / Page 126
—

Thomas – The Trend Factory
Rosenthal AG
Philip-Rosenthal-Platz 1
D-95100 Selb
T +49 92 87 7 25 66
F +49 92 87 7 22 71
beate.reichel@rosenthal.de
www.rosenthal.de
Seite / Page 142
—

Trilux-Lenze GmbH + Co. KG
Heidestraße
D-59759 Arnsberg
T +49 29 31 30 10
F +49 29 31 30 13 75
dpietz@trilux.de
www.trilux.de
Seite / Page 170
—

Tupperware Belgium N.V.
Pierre Corneliskaai 35
B-9300 Aalst
T +32 53 72 72 11
F +32 53 72 72 10
Seite / Page 100, 102
—

Tupperware Deutschland GmbH
Praunheimer Landstraße 70
D-60488 Frankfurt
T +49 69 76 80 20
F +49 69 76 80 22 99
kontakt@tupperware.de
www.tupperware.de
Seite / Page 100, 102
—

Adressen
Addresses

Twin Diver AG
Oderstraße 1
D-89231 Neu-Ulm
T +49 7 31 9 80 75 55
F +49 7 31 9 80 75 56
Seite / Page 194
—

UN Studio
Ben van Berkel
Stadhouderskade 113
NL-1073 AX Amsterdam
T +31 20 5 70 20 40
F +31 20 5 70 20 41
info@unstudio.com
www.unstudio.com
Seite / Page 134
—

Viessmann Werke GmbH & Co. KG
Viessmannstraße 1
D-35107 Allendorf/Eder
T +49 64 52 7 00
F +49 64 52 70 27 80
wee@viessmann.com
www.viessmann.com
Seite / Page 28, 36
—

Vitamin D Design Services
Treppenstraße 17
D-42115 Wuppertal
T +49 2 02 4 46 98 00
F +49 2 02 4 46 98 01
office@vitamin-d.de
www.vitamin-d.de
Seite / Page 98
—

Vitra GmbH
Charles-Eames-Straße 2
D-79576 Weil am Rhein
T +49 76 21 70 20
F +49 76 21 7 02 32 42
info@vitra.com
www.vitra.com
Seite / Page 130
—

Vogt & Partner
Pionierstraße 12
CH-8400 Winterthur
T +41 52 2 08 99 88
F +41 52 2 08 99 89
www.lichtgestaltung.ch
Seite / Page 164
—

Voith Paper GmbH & Co. KG
St. Pöltener Straße 43
D- 89522 Heidenheim
T +49 73 21 3 70
F +49 73 21 37 72 62
info.voithpaper@voith.com
www.voith.de
www.voithpaper.com
Seite / Page 54
—

Walter Knoll AG & Co. KG
Bahnhofstraße 25
D-71083 Herrenberg
T +49 70 32 20 80
F +49 70 32 20 82 54
info@walterknoll.de
www.walterknoll.de
Seite / Page 132, 134
—

Weinberg & Ruf
Ludwigstraße 8
D-70794 Filderstadt
T +49 7 11 7 08 50 10
F +49 7 11 7 08 50 18
info@weinberg-ruf.de
www.weinberg-ruf.de
Seite / Page 48
—

Hannes Wettstein
Zed AG
Seefeldstraße 303
CH-8008 Zürich
T +41 44 4 21 22 22
F +41 44 4 21 22 66
info@zednetwork.com
www.zednetwork.com
Seite / Page 150
—

Wiha Werkzeuge GmbH
Obertalstraße 3–7
D-78136 Schonach
T +49 77 22 95 92 10
F +49 77 22 95 91 60
volker.lehmann@wiha.com
www.wiha.com
Seite / Page 46
—

Wilkhahn
Wilkening + Hahne GmbH + Co. KG
Fritz-Hahne-Straße 8
D-31848 Bad Münder
T +49 50 42 99 93 92
F +49 50 42 99 92 30
info@wilkhahn.de
www.wilkhahn.de
Seite / Page 126
—

Wilkhahn
Wilkening + Hahne GmbH + Co. KG
Vertrieb
Fritz-Hahne-Straße 8
D-31848 Bad Münder
T +49 50 42 99 90
F +49 50 42 99 92 26
Seite / Page 126
—

Wilo AG
Nortkirchenstraße 100
D-44263 Dortmund
T +49 2 31 41 02 75 32
F +49 2 31 41 02 74 80
holger.krasmann@wilo.de
www.wilo.de
Seite / Page 38
—

Karsten Winkels
Leibnizstraße 8a
D-44147 Dortmund
T +49 2 31 7 28 29 98
F +49 2 31 7 28 29 82
www.bauwerkstadt.net
Seite / Page 162
—

WMF
Württembergische Metallwarenfabrik AG
Eberhardstraße
D-73312 Geislingen
T +49 73 31 2 51
F +49 73 31 4 53 87
info@wmf.de
www.wmf.de
Seite / Page 84, 146
—

Wodtke GmbH
Rittweg 55–57
D-72070 Tübingen-Hirschau
T +49 70 71 7 00 30
F +49 70 71 70 03 50
info@wodtke.com
www.wodtke.com
Seite / Page 32
—

Grit Andrea Wolf
Kleinkröbitz 7
D-07751 Milda
T +49 3 64 22 2 23 25
wolf_grit@hotmail.com
Seite / Page 227
—

Würth Solergy
Würth Elektronik GmbH & Co. KG
Reinhold-Würth-Straße 4
D-71672 Marbach am Neckar
T +49 71 44 94 14 26
F +49 71 44 94 14 29
rolf.schmitt@we-online.de
www.we-online.de
Seite / Page 26, 168, 190
—

Cairn Young
Clements Yard Iliffe Street
GB London SE 17 3LJ
T +44 2 07 7 93 81 98
F +44 2 07 5 82 44 89
cairn@btconnect.com
Seite / Page 106
—

Ufuk Yüksel
Kolpingstraße 4
D-84076 Pfeffenhausen
T +49 87 82 10 11
ufuk.yueksel@web.de
Seite / Page 228
—

YXLON International X-Ray GmbH
Essener Bogen 15
D-22419 Hamburg
T +49 40 52 72 94 01
F +49 40 52 72 92 72
bettina.ihlenfeldt@hbg.yxlon.com
www.yxlon.com
Seite / Page 50
—

Zumtobel Licht GmbH
Grevenmarschstraße 74–78
D-32657 Lemgo
T +49 52 61 2 12 42 95
F +49 52 61 2 12 81 42 95
patrick.schwegler@zumtobelstaff.de
www.zumtobelstaff.com
Seite / Page 174
—

Zumtobel Lighting GmbH
Schweizer Straße 30
A-6850 Dornbirn
T +43 55 72 39 06 88
petschulat@zumtobelstaff.co.at
www.zumtobelstaff.com
Seite / Page 172
—

Zumtobel Staff Vertriebs GmbH
Zettachring 6
D-70567 Stuttgart
T +49 7 11 7 27 22 10
F +49 7 11 72 72 21 21
Seite / Page 172
—

Zumtobel Staff Vertriebs GmbH
Grevenmarschstraße 74–78
D-32657 Lemgo
T +49 52 61 21 20
F +49 52 61 2 12 77 77
www.zumtobelstaff.de
Seite / Page 174
—

Zweibrüder Optoelectronics GmbH
Kronenstraße 5–7
D-42699 Solingen
T +49 2 12 5 94 80
F +49 2 12 5 94 82 00
info@zweibrueder.com
www.zweibrueder.com

Werksdesign / In-house design
Wojtek Niemiec
Seite / Page 160
und
Rainer Opolka
Seite / Page 156, 158
—

Namenregister
Index of names

Die fettgedruckten Seitenzahlen verweisen auf den Katalogteil, die mager gedruckten Seitenzahlen führen zum Adressenverzeichnis.
—
Page numbers in bold type refer to the catalogue section, while those in lightface type refer to the index of addresses.
—

Impressum
Publishing details

Herausgeber / Editor
Design Center Stuttgart
Regierungspräsidium Stuttgart
Willi-Bleicher-Straße 19
D-70174 Stuttgart
T +49 7 11 1 23 26 84
design@rps.bwl.de
www.design-center.de
—

Text und Redaktion / Text and editorial supervision
Andrea Scholtz M.A.
büro wortgewandt, Stuttgart
—
Übersetzung / Translation
Gil Webster, Stuttgart
Philip Mann, Marbach
Bruce Stout, Grafenau
—
Grafikdesign, Lichtillustrationen / Graphic design, light illustrations
stapelberg&fritz, Stuttgart
—
Lithografie / Lithography
ctrl-s prepress GmbH, Stuttgart
—
Druck / Printing
Leibfarth + Schwarz GmbH + Co. KG
Dettingen/Erms
—
Produktion / Production
Meike Pätzold
avcommunication GmbH Ludwigsburg
—

Verlag und Vertrieb / Publishing and Distribution
avedition GmbH
Königsallee 57
D-71638 Ludwigsburg
T +49 71 41 1 47 73 91
kontakt@avedition.de
www.avedition.de
—

ISBN-10: 3-89986-074-8
ISBN-13: 978-3-89986-074-0
—
Printed in Germany
—

Die Publikation erscheint anlässlich der Ausstellung »Focus Energy – Internationaler Designpreis Baden-Württemberg 2006 und Mia Seeger Preis 2006« 21. Oktober bis 10. Dezember 2006
—
The publication is published on the occasion of the exhibition »Focus Energy—Baden-Württemberg International Design Award 2006 and Mia Seeger Prize 2006« 21 October to 10 December 2006
—

Veranstalter / Organizer
Design Center Stuttgart
Willi-Bleicher-Straße 19
D-70174 Stuttgart
T +49 7 11 1 23 26 84
—
Projektleitung / Project Head
Sabine Lenk
—
Organisation / Administration
Hildegard Hild
Michael Kern
—
Ausstellungsgestaltung / Exhibition Design
design hoch drei, Stuttgart
—
Inszenierung Preisverleihung / Production of award-giving ceremony
pulsmacher, Ludwigsburg
—